Teaching Plant Anatomy
THROUGH CREATIVE LABORATORY EXERCISES

Dr. R. Larry Peterson
University of Guelph

Dr. Carol A. Peterson
University of Waterloo

Lewis Melville
University of Guelph

Design and Layout
by Forrest Phillips

© 2008 National Research Council of Canada

All rights reserved. No part of this publication may be reproduced in a retrieval system, or transmitted by any means, electronic, mechanical, photocopying, recording or otherwise, without the prior written permission of the National Research Council of Canada, Ottawa, ON K1A 0R6, Canada.

Printed in Canada on acid-free paper.

ISBN 978-0-660-19798-2
NRC No. 49723

Library and Archives Canada Cataloguing in Publication

Peterson, R. Larry

Teaching plant anatomy through creative laboratory exercises / by R. Larry Peterson, Carol A. Peterson and Lewis H. Melville.

Issued by the National Research Council Canada.
Includes bibliographical references.
ISBN 978-0-660-19798-2

1. Plant anatomy--Study and teaching (Higher)--Activity programs.
2. Plant anatomy--Study and teaching (Secondary)--Activity programs.
I. Peterson, Carol Anne II. Melville, Lewis H. III. National Research Council Canada IV. Title.

QK641.P47 2008   571.3'2071   C2008-980108-3

NRC Monograph Publishing Program
Editor: P.B. Cavers (University of Western Ontario)

Editorial Board: W.G.E. Caldwell, OC, FRSC (University of Western Ontario); M.E. Cannon, FCAE, FRSC (University of Calgary); K.G. Davey, OC, FRSC (York University); M.M. Ferguson (University of Guelph); S. Gubins (*Annual Reviews*); B.K. Hall, FRSC (Dalhousie University); W.H. Lewis (Washington University); A.W. May, OC (Memorial University of Newfoundland); B.P. Dancik, *Editor-in-Chief*, NRC Research Press (University of Alberta)

Inquiries: Monograph Publishing Program, NRC Research Press, National Research Council of Canada, Ottawa, Ontario K1A 0R6, Canada.
Web site: http://pubs.nrc-cnrc.gc.ca

Correct citation for this publication: Peterson, R.L., Peterson, C.A., and Melville, L.H. 2008. Teaching plant anatomy through creative laboratory exercises. NRC Press, Ottawa, Ontario. 164 pp.

This book is dedicated to Dr. Elizabeth G. Cutter who, through her enthusiasm and 'hands-on' approach to teaching plant structure, showed the senior author that students can be motivated and enjoy plant anatomy.

TABLE OF CONTENTS

PREFACE viii

INTRODUCTION ix

CHAPTER 1. TYPES OF MICROSCOPES AND MICROSCOPY 1
   Magnification and scale chart 2
   Getting to know a light microscope 3

CHAPTER 2. SECTIONING AND STAINING 5
   Section types 5
   Sectioning methods 5
   Step-by-step instructions for freehand sectioning 6
      Sectioning without backing material 6
      Sectioning large specimens 7
      Sectioning with support material 8
   Staining sections 8
      Staining using a Petri dish 8
      Construction of section holders for staining 10
      Possible problems when staining sections 11

CHAPTER 3. CELL ORGANELLES AND ERGASTIC SUBSTANCES 13
   Observation of intact plant cells 13
   BOX 1: Optical sectioning 14
   BOX 2: Plastids 18
   Chloroplasts and chromoplasts 18
   Amyloplasts 20
   BOX 3: Polarizing microscopy 22
   Other storage reserves 23
   Lipid staining 23
   Protein staining 23
   Tissue printing for protein 24
   BOX 4: Tissue printing 24
   Ergastic substances 25
   Calcium oxalate crystals 25
   Calcium carbonate crystals 26
   Vacuolar pigments 27
   Tannins; polyphenols 27

CHAPTER 4. CELL TYPES IN SIMPLE TISSUES 31
   Parenchyma 31
   BOX 5: Staining hand sections with Toluidine Blue O (TBO) 32
   BOX 6: Parenchyma 33
   Collenchyma 33
   BOX 7: Collenchyma 35

Sclerenchyma	36
Fibres	36
BOX 8: Sclerenchyma	**37**
Sclereids	38
BOX 9: Maceration technique	**39**

CHAPTER 5. COMPLEX TISSUES — 41

Xylem	41
BOX 10: Xylem	**43**
BOX 11: Comparison of tracheary elements of elongating and non-elongating sunflower stem internodes	**44**
BOX 12: Sample key to macerated xylem	**47**
BOX 13: Translocation through the xylem	**48**
Phloem	51
BOX 14: Phloem	**52**

CHAPTER 6. SECRETORY STRUCTURES — 55

External secretory structures	55
Glandular (secretory) trichomes	55
Tissue printing of whole leaves to demonstrate secretory trichomes	56
Nectaries	57
Hydathodes	58
Internal secretory structures	58
Secretory ducts and cavities	58
Laticifers	61

CHAPTER 7. ROOTS — 63

Root caps	63
Root hairs	63
BOX 15: Growing grass seedlings for root hairs and lateral roots	**64**
Primary tissues in roots	65
Exodermis and endodermis	67
BOX 16: Endodermis	**70**
BOX 17: Exodermis	**71**
Roots with phi thickenings	72
Lateral roots	73
Secondary growth in roots	74
BOX 18: Clearing and staining for lateral root primordia	**76**
Specialized roots	76
Aerial roots of orchids	76
Storage roots	78
Roots of aquatic plants	79

CHAPTER 8. STEMS — 81

Shoot apex	81
Primary tissues in stems	83
Secondary tissues in stems	89
Periderm	91

Specialized stems ... 93
Stems of aquatic plants ... 93
Stems with modified secondary growth ... 95

CHAPTER 9. LEAVES ... 99
Epidermis ... 99
BOX 19: Method for making epidermal peels ... 100
BOX 20: Peltate scales on bromeliad leaves ... 101
Covering trichomes ... 102
BOX 21: Clearing leaves ... 104
BOX 22: Simple clearing method for most leaves (and flat floral organs) ... 105
Tissue organization in leaves ... 106
BOX 23: Demonstrating starch in bundle sheath cells of C_4 plants ... 112
Vascular tissues ... 113
Leaves of aquatic plants ... 115
Petiole anatomy ... 117

CHAPTER 10. REPRODUCTIVE ORGANS ... 119
Flowers ... 119
BOX 24: Demonstration of germinating pollen grains ... 129
Embryos ... 130
BOX 25: Clearing Arabidopsis ovules with Hoyer's solution ... 131

APPENDICES ... 132
APPENDIX 1: List of plant anatomy texts ... 132
APPENDIX 2: Preparation and use of stains ... 132
APPENDIX 3: Some unwelcome intruders and problems when sectioning ... 134
APPENDIX 4: A simple method for staining cleared roots and leaves ... 137
APPENDIX 5: Testing viability of plant cells ... 138
APPENDIX 6: Suppliers ... 140

GLOSSARY ... 141

INDEX ... 150

PREFACE

This book has been developed from the authors' extensive experience teaching plant anatomy and their conviction that a 'hands-on' approach is one of the most effective ways to engage students in the learning process. The book is designed as a guide for teaching plant structure in high school, college, and university and is based on students preparing their own slides from fresh plant samples. The exercises included have all been tested and most require minimal supplies and equipment. The plant species used for the various exercises are those that are readily available by growing from seed, collecting in natural environments, or purchasing from grocery stores, greenhouses, and nurseries. Although there is an obvious bias towards the use of plants occurring in the north temperate region of the world, exercises can be adapted easily to plants occurring in other regions.

Students enrolled in a third year course in plant anatomy during the fall semester, in both 2003 and 2004, in the Department of Botany (now the Department of Integrative Biology) at the University of Guelph, have tested most of the methods included in the book. We thank them for their input and their permission to use some of their images. Credit is given in the figure captions for images that are not our own.

No attempt is made to include all aspects of plant structure or the development of cells, tissues and organs; this is left to plant anatomy texts (**see Appendix 1**). Also, we have avoided commercially available microscope slides that are used in the teaching of plant structure since these images can be readily found on websites, in texts, or in laboratory manuals.

We thank the Universities of Guelph and Waterloo for giving us the opportunity to develop courses in plant anatomy that emphasize a 'hands-on' approach to learning. We also thank the Natural Sciences and Engineering Research Council of Canada for their indirect support in providing funds for some of the equipment used in capturing images.

We thank Cameron Wagg, Ryan Geil, Trevor Wilson, and Daryl Enstone for their help with some of the methods. Daryl Enstone and Chris Meyer provided a critical review of the text and contributed some of the images. We thank Ron Dutton, Department of Plant Agriculture, University of Guelph, for allowing us access to plants in the teaching collection, and Denise McClellan and Carole Ann Lacroix for providing plant material. We are indebted to Forrest Phillips for his skill in formatting and layout of the textbook and for his many suggestions that improved the final copy. We thank Suzanne Kettley at NRC Research Press for her careful editing and for seeing the project to completion.

About the authors

Dr. R. Larry Peterson is a University Professor Emeritus at the University of Guelph. He received B.Ed. and M.Sc. degrees from the University of Alberta and a PhD. from the University of California (Davis). He has developed and taught many courses in botany, including plant anatomy. He has an international reputation for his research on symbiotic associations between fungi and roots and is a co-author of a recent book on this topic. He has received several awards for teaching and research. He is a Fellow of the Royal Society of Canada.

Dr. Carol A. Peterson is a Distinguished Professor Emerita at the University of Waterloo. She received a B.Sc. and M.Sc. from the University of Alberta and a PhD. from the University of California (Davis). She has also developed and taught courses in botany, including plant anatomy. She has an international reputation for her research on the structure and function of plant roots, and receives many invitations to participate in international symposia. In recognition of her major contributions to the Canadian Society of Plant Physiologists, she received the Gleb Krotkov Award.

Lewis Melville is a Research Associate in Dr. R. L. Peterson's Research Laboratory. He received a B.Sc. from the University of Guelph and has contributed to research and teaching for over twenty years. He is the co-author of numerous research publications and of a book on mycorrhizas.

INTRODUCTION

How to use this book

The main objective of this book is to provide students, teachers, and researchers with simple methods to investigate the structure of plant cells, tissues, and organs using fresh material and a minimum of supplies. To accomplish this, exercises are included that use plant species that are readily available. For many of the exercises, substitutions can be made if the particular plant species mentioned is not available. Both scientific and common names are given for all plant species used in the book.

Chapter 1 describes microscopes and various microscopical methods. Since most of the exercises require basic compound light microscopes, step-by-step instructions of how to set up a compound microscope for optimal results are outlined. A magnification chart showing the range of magnifications that can be obtained with each type of microscope and the relative sizes of structures observed is included. A few exercises requiring more sophisticated microscopes or more advanced methods are described for those instructors who have access to this technology and want to explore plant structure in more detail.

This is a hands-on book; it is important that the methods outlined for sectioning and staining of samples are followed carefully to obtain satisfactory results. We have, therefore, included detailed instructions for these methods in **Chapter 2**. Problems that might be encountered when preparing samples for microscopy are discussed in **Appendix 3**.

The book includes exercises on cell structure, tissues, and organs. Photomicrographs, with accompanying figure legends, are provided to illustrate the results that can be expected from simple freehand sectioning and staining. In some instances, low magnification and higher magnification images of the same tissue are provided to clarify the location of particular cell types. In a few cases, two stains are used with the same tissue to verify the chemical composition of cell walls. Scale bars are provided for images to indicate the range in size of the structures within the plant body. All of the images are included on the accompanying CD-ROM and can be downloaded into PowerPoint presentations. In some cases, labeled diagrams are included to illustrate more difficult concepts and as aids in understanding particular structures. The stains used in the book are listed in **Appendix 2** along with details of their preparation and the protocols for staining.

A number of single page 'boxes' that contain topics of special interest are found in each chapter. A glossary is included with definitions for most of the terms used throughout the manual. Terms in the glossary are bolded when first mentioned in the text. An index is provided to aid the reader in locating particular topics. Appendices relative to topics covered in the book, as well as a list of suppliers of scientific equipment and supplies, are also included.

This book can be used either as a laboratory manual or as a resource for students and instructors in university courses dealing primarily with plant structure. However, the book is organized so that individual exercises can be adapted for courses in high school, college, and university curricula in which plant structure is only a minor component of the content.

We view this book as a starting point for students, teachers, university instructors, and researchers in the exploration of plant structure. Once the basic methods of sample preparation (usually freehand sectioning and staining) have been mastered they can be applied to project-based courses and research in plant structure, as well as providing correlative structural information for studies in plant physiology and molecular biology.

Chapter 1

Types of microscopes and microscopy used in this book

Stereo binocular or dissecting microscope

Binocular microscopes allow the user to view the subject magnified up to 250 times its normal size; they are used primarily to study the morphology of a specimen. The magnification is relatively low, however, and it is difficult to distinguish structure on a cellular level; sub-cellular detail is not possible. At high magnification, most of the material appears out of focus because of the low depth of field.

Compound microscope (transmitted light microscope)

With this microscope, light is transmitted (usually from below) through a specimen that is mounted on a glass slide, and then focused through a series of lenses. Special lenses and optical refinements make magnifications of up to 2000 times possible. While most material is usually sectioned, mounted on a slide, and then stained, it is possible to observe some living specimens if they are thin enough to allow the passage of light. Cell types, many of the larger cellular organelles (plastids, nuclei, vacuoles), and metabolic products such as calcium oxalate crystals and starch granules can be distinguished. Since this is the type of microscope to be used throughout the book, details of how to get the best images are provided.

Nomarski interference contrast microscopy and differential interference contrast microscopy

These types of microscopy amplify the small differences in refractive index of cellular components in unstained or weakly stained specimens. They are extremely useful for revealing cytological details of living cells. Special Nomarski filters and objective lenses are substituted for the regular lenses on a compound microscope. When polarized light passes through a Wollaston prism it splits the incident light into two parallel beams, which then pass through the specimen. If one of the beams is diffracted by material within the specimen, an interference pattern is formed when the two beams are recombined by a second prism above the specimen. The birefringent effect at the edges of different structures in the specimen increases the contrast and definition of structures such as vacuoles, cytoplasmic strands, and nuclei that are otherwise difficult to see in a normal compound microscope. Heavy staining reduces the effect.

Epifluorescence microscopy

Fluorescence microscopy is an important method in research and in the teaching of plant structure. A main value of the technique is its ability to reveal substances at very low concentrations.

Fluorescence is an optical phenomenon in which light of a short wavelength is absorbed by a substance and re-emitted as a light of a longer wavelength. In epifluorescence microscopy, a compound microscope is modified so that the light illuminates the specimen from above. The light usually is in the blue to ultraviolet range and is provided by a suitable excitation filter in combination with a fairly intense light source, usually a mercury vapour lamp. These short wavelengths of light are screened from entering the eye by a barrier filter, so that only the longer, visible wavelengths of light are detectable.

Many substances occurring in plant organs show autofluorescence (primary fluorescence) when illuminated by light with a short wavelength. For example, cutin, suberin, lignin, many phenolic compounds, and chlorophyll show characteristic fluorescence. Hand sections of fresh tissue can be used to demonstrate lignified elements, cuticle, suberized walls of phellem cells, Casparian strips of endodermal cells, and phenolic deposits within cells. Also, there are many fluorescent probes available for the identification of cellular organelles.

Polarizing microscopy

See Box 3.

Magnification and Scale Chart

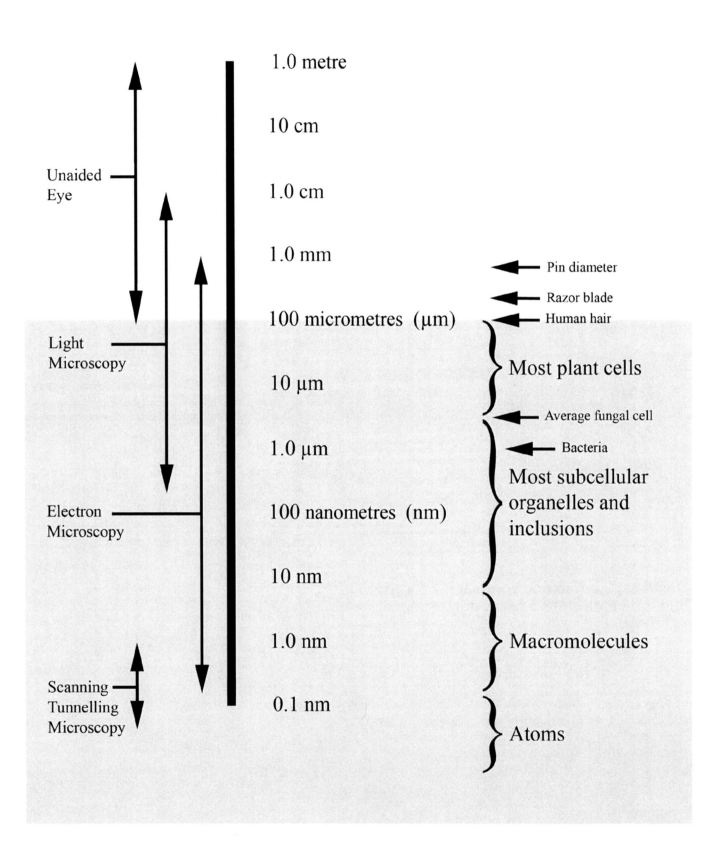

Getting to know a light microscope

Parts of the compound microscope

The following diagram illustrates the parts of a typical binocular compound microscope.

Beginning students should familiarize themselves with the locations and functions of the following parts of a microscope.

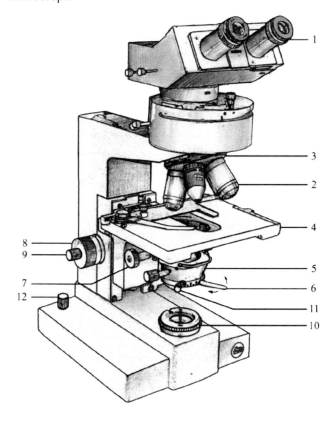

1. Eyepieces. Adjustable focus on one or both to compensate for eyesight differences between individual users.
2. Objective lenses. There may be several of these. Usually 4×, 10× and 40× are included as a minimum.
3. Nosepiece. The base that supports the objectives. The nosepiece can be rotated to view the specimen through different objective lenses.
4. Microscope stage and the device for holding a slide in place. This device may consist of simple clips, or an apparatus that can be moved with the aid of screws below or above the stage.
5. Condenser. This is part of the illuminating system consisting of one or more lenses.
6. Iris (aperture) diaphragm and the lever or ring whereby it can be opened and closed.
7. Condenser adjustment knob.
8. Coarse focus knob that moves the stage through long distances rapidly.
9. Fine focus knob that moves the stage through short distances more slowly.
10. Field diaphragm.
11. Condenser mount with sledge guide and centering screws. The light path through the condenser can be aligned using these screws, so that an even field of illumination is achieved.
12. Light intensity control.

It is recommended that you consult the appropriate manual before using a microscope.

When not in use, microscopes should be kept covered. After using wet mounts, it is important to clean the objective lenses to remove mounting solution that may have been left there inadvertently. Dirty lenses should be cleaned with a drop of lens cleaner on a piece of lens tissue. Do not use other types of tissue, as the lens may be scratched. Then, polish the lens using a clean, dry portion of lens tissue and a circular motion with light pressure.

Basic instructions for microscope use

These instructions are for a binocular microscope. However, if a monocular microscope is being used, omit steps 8 and 9.

1. Plug in the microscope.
2. Turn on the light.
3. Place the slide, cover glass uppermost, on the microscope stage.
4. Rotate the 4× (or 10×) objective lens into place.
5. Adjust the light intensity so that it is comfortable for viewing.
6. While looking at the slide from the side, raise the stage as far as possible with the coarse focus knob. Never raise the stage with the coarse focus knob while looking through the eyepieces.
7. Looking through one eyepiece only, slowly lower the stage with the coarse adjustment knob until the image of the specimen comes into focus.
8. Adjust the interpupillary distance for your eyes. While viewing the specimen with both eyes, push the eyepieces together or pull them apart until one image is seen.
9. Look through each eyepiece in turn and test whether or not it can be focused individually by rotating the focusing ring near the top of the eyepiece while observing the specimen. If only one eyepiece can be focused, use the other eye and eyepiece to focus on an object within the specimen. Then, without touching the focusing knobs of the microscope, look at the specimen with the first eye and eyepiece, and bring the object into focus by turning the eyepiece focusing ring. If both eyepieces can be focused, this step can be done in any order.
10. To achieve a greater magnification of the specimen, rotate the 10× (or 40×) objective into place.

11. Use the fine focus knob to adjust the focus if necessary.
12. If a higher magnification is desired, carefully swing the 40× objective into place.
13. Adjust the light intensity for each specimen at each magnification.
14. Adjust the condenser by raising or lowering it using the condenser adjustment knob, until the most even illumination for each specimen and magnification is obtained. Adjusting the condenser will also achieve maximum contrast. The field diagram should be opened so the light just fills the field of view.
15. To remove the slide, rotate the 40× objective so that either the 4× or 10× or no objective is in place. Only then remove the slide from the stage.

Note: Be sure to include steps 13 and 14 for each specimen as they are important for obtaining satisfactory images.

Chapter 2
Sectioning and staining

Section types

Although it is possible to observe a few plant parts directly, for example with a stereo binocular microscope, usually the cells of interest are located within the plant body. Sectioning is required to allow sufficient light to pass through a specimen when using a compound microscope. Most students learn to make good sections after a few hours of practice.

Various types of sections can be prepared depending on the shape of the organ. In the diagrams below and to the upper right, the orientation within the organ is indicated by shading. Shaded sections to the right of each diagram show its shape as it would appear when mounted on a microscope slide.

Sectioning methods

When making cross (transverse) sections, it is important to cut at right angles to the longitudinal direction of the stem (or other organ) as illustrated below and in figure A to the lower right. This can be judged by inspecting the cut end of the stem at intervals during sectioning.

In cutting longitudinal sections, it is helpful to make a partial cut into the organ a few millimetres from the severed end as illustrated in figure B to the lower right, and section from this end to the partial cut.

In flat organs such as leaves and petals, three types of sections are possible:

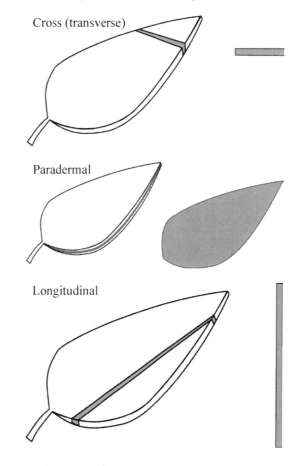

Cross (transverse)

Paradermal

Longitudinal

In columnar organs such as stems and roots, the following types of sections can be made:

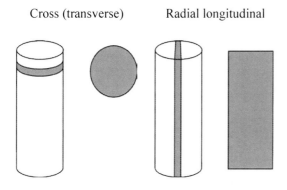

Cross (transverse)   Radial longitudinal

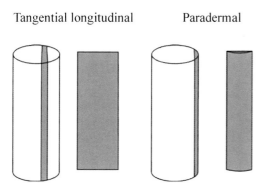

Tangential longitudinal   Paradermal

Cross section

Longitudinal section

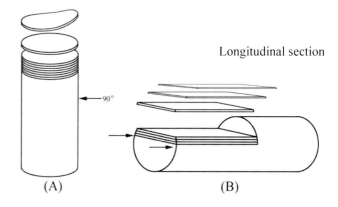

(A)   (B)

Step-by-step instructions for sectioning

When fresh material is sectioned, some cells are sliced open and their contents are released. These can be removed by rinsing the sections in water in a Petri dish before mounting them on a slide for observation. Very thick sections will not allow enough light to pass through to reveal cellular detail, and very thin sections will result in the removal of most cellular contents. The desired thickness, therefore, lies somewhere between these two extremes. Partial sections (in which part of the specimen is not present) are often very good to observe because a gradient in thickness is present in a single section.

Sectioning without backing material

The following instructions are illustrated with a stem.

1. Assemble the required equipment in an area with good lighting where the person can be seated:
   - Slides and cover glasses
   - Beaker filled with water for dipping (not shown)
   - Small paintbrush/probes
   - Dropper bottle (or flask) with water
   - Small Petri dishes
   - New, **double-sided** razor blades (**Snap in half before removing the paper covering.**)
   - Plant material to be sectioned
2. Either place a drop of water on a slide or fill the bottom of a small Petri dish with water. If a slide is used, always put the mounting medium on the slide first and add the sections to it to minimize the creation of air bubbles.
3. Cut the part to be sectioned from the plant and place the cut end in a beaker of water. Return cut plant part to water when it is not being sectioned.
4. Dip the razor blade in water and make a leveling cut for the purpose of forming a right angle with the axis of the organ. This will ensure that the sections made will be cross sections and not oblique sections. Whenever making sections, make sure both the specimen and the blade are wet. There are two motions that can be made when sectioning.
5. Grip the stem near the area to be sectioned between the thumb and index finger.
6. Steady hands by resting both elbows on the table.
7. With the razor blade in the other hand, relax and cut 6–10 sections from the end of the stem, either by drawing the razor blade across it or by making back-and-forth sawing motions (see diagrams). Most of the cut sections will collect on the blade. Only a fraction of the sections will be suitable

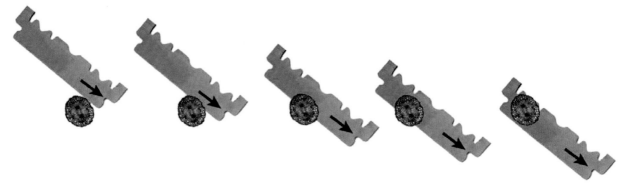

The preferred method is to make cross sections with a smooth, oblique motion.

Sectioning and Staining

Cross section being taken with a sawing motion.

for viewing; thus, it is necessary to make many sections to ensure getting some thin ones. Sectioning plant material dulls razor blades quickly, and only a few sections can be made with one area of the blade. Move to a new area of the blade frequently.

8. At intervals, check the angle of sectioning by holding up the stem and looking at the cut edge. The cut face should be at a right angle to the axis of the stem. If this is not the case, adjust the position of either your hand or the razor blade so as to achieve sections at the proper angle. Wet the severed end and razor blade, and continue sectioning.

9. With a wet paintbrush or probe, remove the sections from the edge of the razor blade and place them either in a drop of water on the microscope slide or into a small Petri dish with water. Inspect the sections and remove any that are too thick. The thinnest ones will be almost transparent. Do not allow sections to dry out.

Sectioning large specimens

When the plant material (e.g., a celery stalk, potato tuber, or carrot root) to be sectioned is large, trim the material to achieve a face no larger than 3 mm × 3 mm.

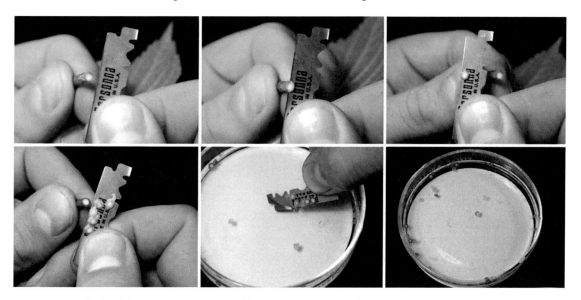

Method for sectioning fairly large plant organs with a two-sided razor blade. In this example, a stem is cut from the plant and held between the thumb and index finger. Sections are prepared by drawing the razor blade over the surface of the stem in a smooth motion with the blade oriented at a right angle to the axis of the specimen. Sections are collected in a small dish containing water. The thinnest sections can either be mounted directly in a drop of mounting medium on a glass slide or be stained before viewing. Transfer sections with a wet paintbrush.

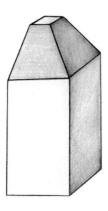

Diagram illustrating a piece of potato (*Solanum tuberosum*) tuber trimmed prior to sectioning.

Sectioning with support material

In cases where the plant tissue is not strong enough to be sectioned with the usual method, it can often be cut when supported by some other material. Leaves and petals usually require support. In the following example, a very small stem is sectioned with a Styrofoam support. If this is not available, some other material with the suitable consistency (such as piece of pickled carrot root) can be substituted. Sectioning using a Styrofoam support is illustrated on the next page.

1. In addition to the materials previously listed (refer to step 1 of **Sectioning without backing material**) you will need the following:
   - Support material, e.g., Styrofoam
   - Fine forceps
2. Trim a piece of Styrofoam so that there is a pyramid at one end leaving an uncut piece at least 1 cm long to hold onto (as in **Sectioning large specimens**, above).
3. Make a longitudinal slit into the Styrofoam with a razor blade.
4. Pry open the slit in the Styrofoam and carefully insert stem piece.
5. Trim off the exposed part of the stem piece.
6. Holding the slit closed, dip the Styrofoam and the enclosed stem in water.
7. Dip the razor blade in water and section across the Styrofoam at right angles to the slit. Pieces of Styrofoam and the sections will collect on the blade.
8. Using a wet paintbrush, slide the material into a water drop on a slide or into a small Petri dish with water. The Styrofoam pieces usually fall away from the plant material and can be removed from the water with forceps.

A similar procedure can be followed for long, thin organs like roots by making a hole instead of a slit in the backing material.

Staining sections

Although sections of some plant organs have enough natural contrast to enable structural details to be determined, it is often necessary to use a stain to provide contrast. There are many stains that can be used and a number have been listed in **Appendix 2**. There are a number of ways in which freehand sections can be stained.

Staining using a Petri dish

Using a small, wet paintbrush, place sections in a small Petri dish (or any flat-bottomed dish) containing the stain of choice. To reduce the amount of stain needed, a drop of the stain can be placed on a piece of Parafilm™ or waxed paper in the bottom of the dish. More than one section can be placed in the stain at a time.

After the recommended time for staining, sections should be picked up with a wet paintbrush and placed in a small dish containing water. They should be agitated with the paintbrush to remove excess stain. Rinsed sections can then be placed in the recommended mounting medium (usually water) on a microscope slide. A cover glass should be added as illustrated.

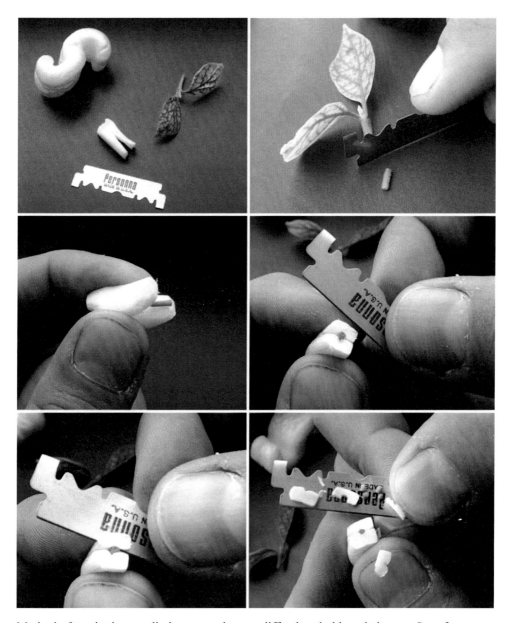

Method of sectioning small plant parts that are difficult to hold on their own. Styrofoam pieces can be used to support the tissue either by holding the tissue between two pieces or by making a small hole or slit in one Styrofoam piece to hold roots or leaves, respectively. Sections prepared in this way can be mounted directly in water under a cover glass or be stained before mounting.

Add a cover glass by lowering it at an angle onto the drop of water containing sections. If needed, add more water or other medium to one edge of the cover glass until the underside of the cover glass is uniformly wet.

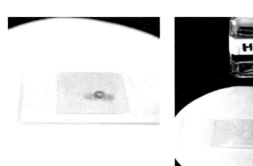

Alternatively, sections can be placed in a drop of stain directly on a slide for the recommended time. Draw stain off with a Kimwipe or filter paper and wash sections with water or other medium by placing a drop on the sections and then drawing it off. Repeat if necessary. Add a drop of water or other medium and then add a cover glass.

Construction of section holders for staining

When a series of staining and rinsing steps need to be conducted with freehand sections, or when many sections need to be processed, it can be advantageous to use section holders. These can be made with a variety of materials and basically consist of a small plastic cylinder with a mesh bottom in which to hold the sections. It is important that the mesh be open enough to allow liquid to pass through, but smaller than the sections to be placed within the holder. The section holders illustrated below have been constructed from BEEM™ capsules. Microcentrifuge tubes can also be used.

Using a razor blade remove and discard the tapered bottom of the original capsule. Similarly, remove and discard the top of the capsule lid, leaving a ring of plastic that will snap over the tube, holding the mesh in place.

A square of nylon mesh (45–50 μm pore diameter) is placed over the open end of the tube and the ring of the lid forced around it to hold it in place. Excess mesh material should be trimmed. Nylon mesh is not compatible with strong acids; do not use the capsules with, for example, phloroglucinol-HCl.

BEEM™ capsule, with removed bottom, top of lid, and remaining ring.

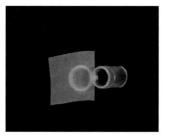 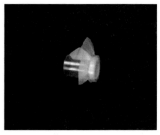

Square of mesh placed between lid ring and the capsule.

Mesh fitted onto the capsule and secured with lid ring.

 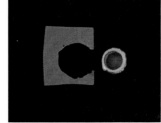

Excess mesh is trimmed.

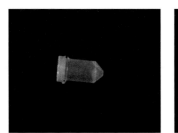

BEEM™ capsules.

A series of small Petri or other flat-bottomed dishes need to be set up to take the sections through the desired staining and rinsing procedure. To begin, the capsules are placed mesh side down in a small Petri dish to which a small amount of water has been added. Freehand sections are transferred from the razor blade to the water in the capsules by means of a small paintbrush.

To begin the staining process, a capsule containing sections is lifted from the Petri dish and the excess water is removed by lowering it onto absorbent paper.

Capsule transferred by forceps.

Water blotted from capsule.

Capsule transferred to staining solution.

Excess staining solution blotted off.

The capsule is then placed in the stain for the desired amount of time. It can be gently agitated to ensure contact between the stain and all sides of the sections.

The capsule is lifted out and the excess stain removed by lowering it onto absorbent paper. Sections may then be rinsed briefly with a gentle stream of water, and, if a longer rinsing time is desired, put into dishes of water (or a series of dishes) for the desired times. At the end of the rinsing, sections can be removed by lifting them out with a paintbrush and placed either in water in a Petri dish or in a drop of water on a microscope slide for viewing.

Sections rinsed with a stream of water.

Capsule containing stained sections in a dish of water.

If desired, sections can be left in the capsule before or after staining and stored in a refrigerator for a few days in a covered Petri dish with water. Alternatively, sections can be stored for longer times in 50% ethanol or other preserving solutions.

Possible problems when staining sections

In cases where sections are stained prior to viewing, several problems can arise. Precipitates may be scattered over the sections and the mounting medium. This can happen when the stain has precipitated in the bottle or when some component of the sections reacts with the stain causing it to precipitate. The latter can be minimized by rinsing the sections before mounting them on the slide. If the former has occurred, the stain can be filtered to remove the precipitate or a fresh batch prepared.

The time of staining may be too long, leading to over-staining, or too short, resulting in under-staining. The optimum time for staining varies with the plant material being used, so a range of times should be tested.

Occasionally, mucilage (slime) produced by some plant cells will interfere with staining of plant tissues.

Chapter 3

Cell organelles and ergastic substances

Observation of intact plant cells

It is possible to observe intact plant cells directly with a compound microscope when the cells normally occur singly or in a thin layer. For example, trichomes (hairs) on leaf and stem surfaces, root hairs, thin leaves, and petals can be examined in this way as the entire cell or group of cells can be mounted directly on a slide.

In viewing intact cells, the thickness of the cell is usually greater than the focal depth of the microscope. Thus, it is possible to focus on only one layer in the cell at any one time. Focusing up and down on the specimen will allow all the details of the cell to be observed. An example of viewing a cell in successive focal planes is illustrated in **Box 1**. It is essential to know how to use a microscope properly in order to achieve optimal results when viewing specimens mounted on microscope slides. The proper use of a compound microscope is outlined in **Chapter 1**.

Flowers of most wandering jew/wandering sailor (*Tradescantia*) species such as *T. pallida* (Fig. 1) have **stamens** with several multicellular **trichomes (hairs)** along their **filaments** (Fig. 2). These trichomes, consisting of a series of cells (Fig. 3), are excellent for viewing the contents of cells as well as **cytoplasmic streaming (cyclosis)**. With forceps, remove a stamen and mount it in water under a cover glass on a microscope slide. The method of mounting fresh specimens on microscope slides is outlined in **Chapter 2**. Be careful when applying the cover glass that the trichomes are not damaged. Locate the hairs with the 10× objective, and then go to a higher power. View one cell at various focal planes.

The **cell wall**, **nucleus**, **vacuole**, and **cytoplasm** should be visible (Fig. 4). Note the movement of small organelles by

Figure 1. Flower of *Tradescantia pallida* showing petals (arrowhead) and stamens (double arrowhead).

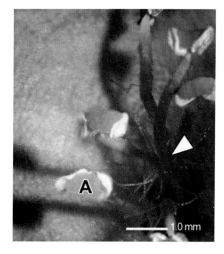

Figure 2. Stamens of *Tradescantia pallida* flower showing anthers (A) and filaments (arrowhead) with trichomes.

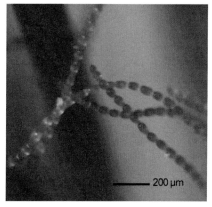

Figure 3. Multicellular trichomes on filaments of a *Tradescantia pallida* flower.

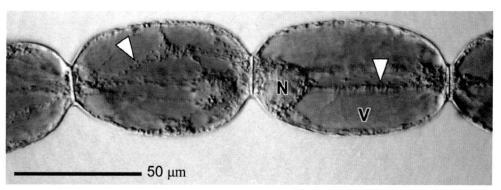

Figure 4. Individual cells of a multicellular trichome from a filament of a *Tradescantia pallida* flower. The nucleus (N), transvacuolar strands of cytoplasm (arrowheads), and vacuoles with anthocyanin pigment (V) are evident. Viewed with Nomarksi interference contrast microscopy.

cytoplasmic streaming within the **transvacuolar strands** of cytoplasm. Focus on the surface of a cell. The ridged structure is the **cuticle** that covers the external surface of the cell wall (Fig. 5).

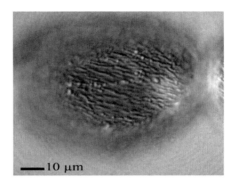

Figure 5. Surface view of a cell from a multicellular trichome of a *Tradescantia pallida* stamen. The outermost part of the wall, the cuticle, is ridged.

Box 1. Optical sectioning

Viewing a cell at one focal plane provides information about only two dimensions of a cell. However, it is possible to determine something about its third dimension by focusing. This is usually done using the 40× or higher magnification objective of a compound microscope. A lens will have a certain depth of field, i.e., a depth at which parts of the specimen are in focus. Generally speaking, the greater the magnification, the smaller the depth of field.

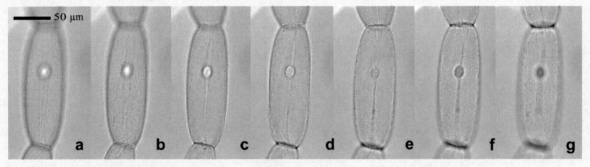

A series of optical sections through an individual staminal trichome cell of *Commelina* sp. Focusing begins at the upper surface of the cell (a) and ends at the lower surface (g). Note that different cellular structures are in focus in this series. The conspicuous granular cytoplasm provides a contrast with the large, clear vacuole in this plant material.

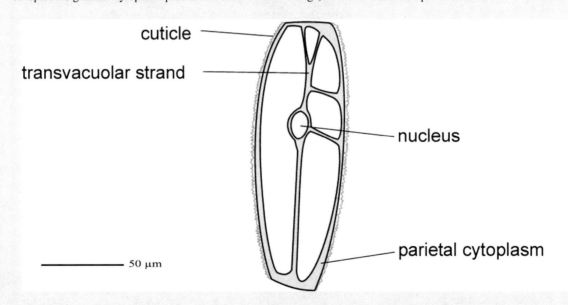

Structure diagram of a staminal trichome cell in *Commelina* sp.

Tomato (*Solanum lycopersicum*) stems, leaf blades, and petioles have two types of trichomes: those that secrete volatile oils and non-secretory 'covering' trichomes, which are illustrated in Fig. 6. The latter are multicellular (Fig. 7) and can be used to demonstrate cytoplasmic streaming. Using fine forceps, detach several covering trichomes at their bases, mount them in water under a cover glass, and locate an intact hair cell using the 10× objective. Using the 40× objective, observe cytoplasmic streaming and the transvacuolar strands in which the streaming is evident (Fig. 8).

Leaves of the Canadian pondweed (waterweed), *Elodea canadensis* (or *Egeria* spp. native to South America), are thin enough to be mounted directly for microscopic observation. With forceps, remove several leaves near the shoot apex, place them in a drop of water on a slide, and place a cover glass over the specimens. Most cells have numerous **chloroplasts** (Fig. 9), which may show movement due to cytoplasmic streaming. This is usually best seen in the narrow cells towards the center of the leaf. To demonstrate **nuclei** in cells containing numerous chloroplasts, leaves can be rinsed briefly in 50% ethanol, stained with the fluorochrome, DAPI (4'-6-diamidino-2-phenylindole; **see Appendix 2**), and viewed with ultraviolet (UV) light using an **epifluorescence microscope** (Fig. 10). Small unicellular trichomes are present along the margin of the leaf; these have fewer chloroplasts and therefore the nuclei, which are almost transparent and are easier to see (Fig. 11).

A simple method of preparing individual plant cells for the study of organelles is to grind **cladophylls** (the leaf-like organs) of the asparagus fern, *Asparagus densiflorus*, in a mortar and pestle containing enough water to wet the tissue. After grinding, take an aliquot of the water that will now appear green and mount it under a cover glass. Individual cells containing chloroplasts will be evident (Fig. 12). When these are viewed with UV light using an epifluorescence microscope, the chloroplasts appear red because of the presence of the **chlorophyll** molecules within them (Fig. 13). Isolated cells can be stained with the fluorescent probe Rhodamine 123 **(see Appendix 2)** and examined with blue light using an epifluorescence microscope to reveal **mitochondria** (Fig. 14).

There are various methods used to test the viability of plant cells; some of these are outlined in **Appendix 5**.

Figure 6. Stem and petioles of tomato (*Solanum lycopersicum*) showing covering trichomes (arrowhead).

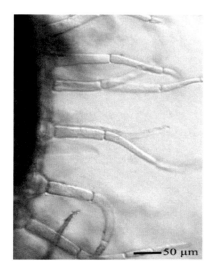

Figure 7. Transverse section of tomato (*Solanum lycopersicum*) petiole showing covering trichomes.

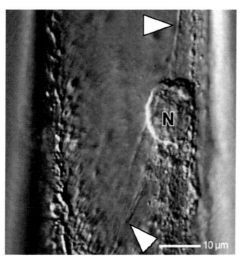

Figure 8. A cell from a tomato (*Solanum lycopersicum*) covering trichome showing the nucleus (N) and transvacuolar strands (arrowheads). Viewed with Nomarski interference contrast microscopy.

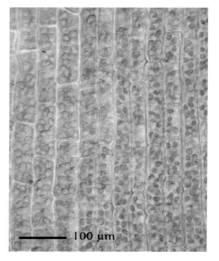

Figure 9. Whole mount of Canadian pondweed (*Elodea canadensis*) leaf showing chloroplasts.

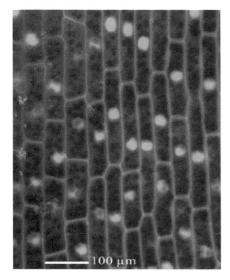
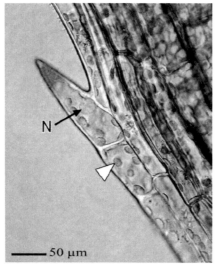
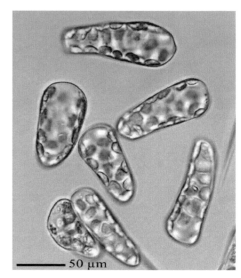

Figure 10. Whole mount of Canadian pondweed (*Elodea canadensis*) leaf stained with DAPI and viewed with UV light showing nuclei.

Figure 11. Whole mount of Canadian pondweed (*Elodea canadensis*) leaf showing chloroplasts and a unicellular trichome with a nucleus (N) and chloroplasts (arrowhead).

Figure 12. Individual cells from cladophylls of asparagus fern (*Asparagus densiflorus*) showing chloroplasts.

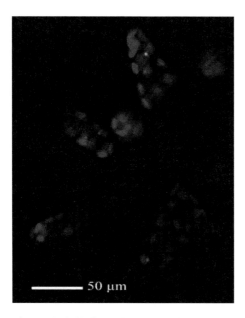
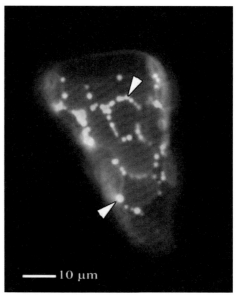

Figure 13. Cells from cladophylls of the aparagus fern (*Asparagus densiflorus*) viewed with UV light using an epifluorescence microscope.
Chloroplasts appear red.
Photo courtesy of Daryl Enstone.

Figure 14. Individual cells from cladophylls of the aparagus fern (*Asparagus densiflorus*) stained with Rhodamine 123 and viewed with blue light using an epifluorescence microscope. Arrowheads indicate mitochondria.
Photo courtesy of Daryl Enstone.

Cell Organelles and Ergastic Substances

A simple method of obtaining individual plant cells for microscopic examination is to use onion (*Allium cepa*) bulbs that can be purchased from a grocery store. These can be sliced (Fig. 15a) and then separated into sections (Fig. 15b) that consist of fleshy scale leaves. The exposed concave surface of a scale leaf can be scored with a sharp razor blade and then pieces of the epidermis can be lifted from the inner surface with a fine pair of forceps (Figs. 15c, 15d). The epidermal peel should then be placed in a drop of water on a microscope slide with its outer surface uppermost (Fig. 15e) and a cover glass added (Figs. 15f, 16). Locate the epidermal cells using the 10× objective and then by using the 40× objective examine individual cells until one showing cytoplasmic streaming is evident (Fig. 17). Observe cytoplasmic streaming and the transvacuolar strands in which this occurs. A nucleus will be obvious within each cell.

Figure 15a. Onion (*Allium cepa*) bulb sliced into sections.

Figure 15b. Scale leaves of an onion (*Allium cepa*) bulb.

Figure 15c. Surface of an onion (*Allium cepa*) bulb scale leaf showing the removal of a portion of the epidermal layer with fine forceps.

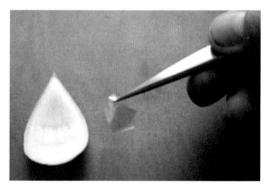

Figure 15d. A transparent epidermal peel from an onion (*Allium cepa*) bulb scale leaf.

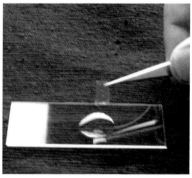

Figure 15e. Onion (*Allium cepa*) epidermal peel being placed in a drop of water on a microscope slide.

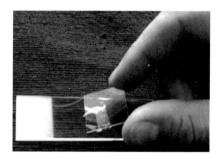

Figure 15f. A cover glass being added to an onion (*Allium cepa*) epidermal peel.

Figure 16. Microscope slide of an onion (*Allium cepa*) epidermal peel ready for observation with a light microscope.

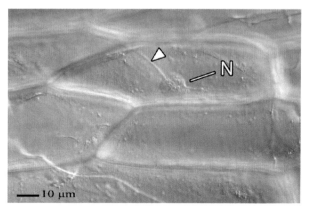

Figure 17. Cells of an onion (*Allium cepa*) epidermal peel showing nuclei (N) and transvacuolar strands (arrowhead).

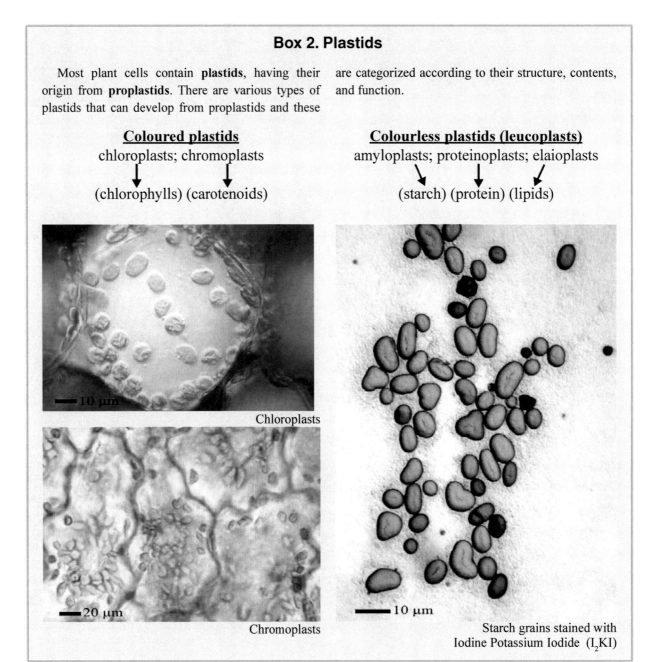

Box 2. Plastids

Most plant cells contain **plastids**, having their origin from **proplastids**. There are various types of plastids that can develop from proplastids and these are categorized according to their structure, contents, and function.

Coloured plastids
chloroplasts; chromoplasts
↓ ↓
(chlorophylls) (carotenoids)

Colourless plastids (leucoplasts)
amyloplasts; proteinoplasts; elaioplasts
↓ ↓ ↓
(starch) (protein) (lipids)

Chloroplasts

Chromoplasts

Starch grains stained with Iodine Potassium Iodide (I_2KI)

Chloroplasts and chromoplasts

Chloroplasts and **chromoplasts** appear coloured because of the presence of various pigments. Chloroplasts contain chlorophylls, among other pigments, giving them their green colour. Chloroplasts can be found in all green tissues in plants. Chromoplasts usually contain yellow and orange **carotenoid pigments** (**carotenes** and **xanthophylls**) and can be found in tissues that are yellow, orange, and sometimes red.

As noted above, *Elodea* (or *Egeria*) leaves as well as cells isolated from asparagus fern cladophylls are suitable for demonstrating chloroplasts but any green stem or leaf can be used.

An interesting exercise is to compare the plastids in a variety of sweet or bell peppers (*Capsicum annuum*). Purchase green, yellow, orange, and red peppers and prepare thin sections of each fruit type with a sharp razor blade **(see Chapter 2)**. Mount these sections in water under a cover glass and observe the plastids.

Green peppers contain many small chloroplasts (Fig. 18), whereas the coloured peppers contain many small chromoplasts (Fig. 19). This exercise can be used to illustrate the conversion of chloroplasts to chromoplasts during fruit ripening. An alternate exercise is to compare the plastids in green versus ripe tomato (*Solanum lycopersicum*) fruits.

Orange and yellow flower petals also contain chromoplasts. Select a variety of coloured petals and prepare them either by sectioning with a two-sided razor blade using a Styrofoam support **(see Chapter 2)** or by tearing them into small pieces. Place in a drop of water on a slide and cover with a cover glass. The edges of torn petals are often thin enough to view the chromoplasts (Figs. 20, 21).

The crystalline nature of the carotenoid pigments in chromoplasts of Bird of Paradise (*Strelitzia reginae*) can be shown by using polarizing microscopy **(Box 3)**. This can be demonstrated by comparing the chromoplasts in Figures 21 and 22.

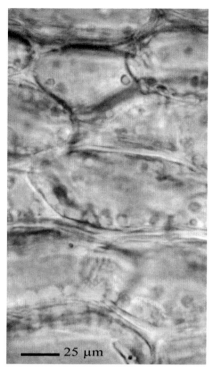

Figure 18. Unstained section of a green fruit of sweet pepper (*Capsicum annuum*) showing chloroplasts.

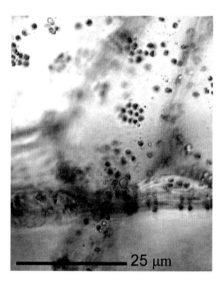

Figure 19. Unstained section of an orange fruit of sweet pepper (*Capsicum annuum*) showing chromoplasts.

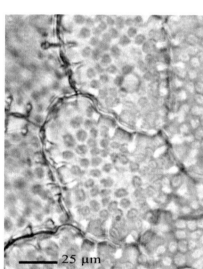

Figure 20. Unstained mount of a petal from a pansy (*Viola odorata*) flower showing chromoplasts.

Figure 21. Unstained section of a petal from a flower of Bird of Paradise (*Strelitzia reginae*) showing elongated crystals in chromoplasts.

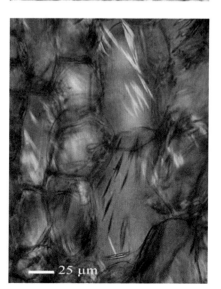

Figure 22. Unstained section of a petal from a flower of Bird of Paradise (*Strelitzia reginae*). Elongated chromoplasts viewed with polarizing microscopy show the crystalline nature of the carotenoid pigments within chromoplasts.

Carrot (*Daucus carota*) roots contain very unusual chromoplasts, since the carotenoid pigment is deposited in the form of a thin, sheet-like crystal that may be curled or twisted. Prepare thin sections of a portion of a carrot root **(see Chapter 2)**, mount in water under a cover glass, and observe the various shaped chromoplasts in most cells (Fig. 23). A simple way to demonstrate the variation in the size and shape of chromoplasts in carrot roots is to press the cut surface of a root on a microscope slide, add a drop of water, and cover with a cover glass. Figure 24 illustrates the diversity in chromoplast size and shape. By using polarizing microscopy **(Box 3)** the crystalline nature of the carotenoid deposits in the chromoplasts can be revealed (Fig. 25).

Amyloplasts

Amyloplasts, along with **proteinoplasts** and **elaioplasts**, are colourless plastids belonging to the broad plastid category, **leucoplasts**. Of the three types, amyloplasts are the most common and are particularly numerous in many storage organs and seeds. Leucoplasts are difficult to demonstrate with light microscopy because they lack pigments. They can be located, however, by staining their contents.

Amyloplasts store starch, and this carbohydrate reserve can be stained with iodine potassium iodide (I_2KI). Potato (*Solanum tuberosum*) tubers are easy to use to demonstrate starch within amyloplasts, since nearly every cell is a storage cell.

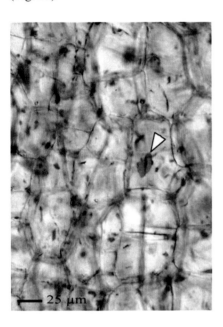

Figure 23. Unstained section of a carrot (*Daucus carota*) root showing variation in size and shape of chromoplasts (arrowheads).

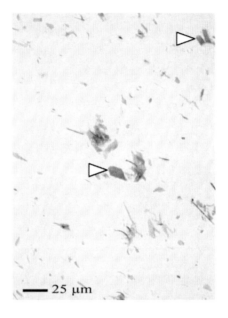

Figure 24. Chromoplasts (arrowheads) of carrot (*Daucus carota*) root that was pressed onto a slide.

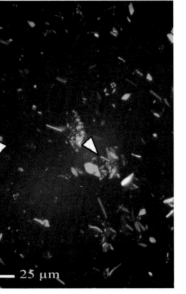

Figure 25. Chromoplasts (arrowheads) of carrot (*Daucus carota*) root that was pressed onto a slide and viewed with polarizing microscopy showing the crystalline nature of the carotenoid pigments within chromoplasts.

Trim a piece of potato tuber **(see Chapter 2)** and cut thin sections, including some of the peel (**periderm**), using a two-sided razor blade. Place these in a drop of I_2KI **(see Appendix 2)** on a slide. Let the tissue stain for a minute or more, draw off the stain with a Kimwipe or filter paper, and mount in a drop of water under a cover glass. Potato tuber cells contain large **simple starch grains** that stain purple with I_2KI (Fig. 26). A component of the starch within the oval starch grains is deposited in a crystalline form, so that when starch grains are viewed with polarizing microscopy **(Box 3)**, a typical Maltese cross configuration is evident. Prepare thin sections of a potato tuber, mount directly in water under a cover glass, and view with polarizing microscopy. Starch grains showing an asymmetrical Maltese cross configuration will be evident (Fig. 27). Starch grains stained with I_2KI do not show the Maltese cross pattern. Although few in number, cuboidal deposits of protein are present in cells next to the peel (Fig. 28).

A simple means of observing starch grains is to cut an imbibed seed of common bean (*Phaseolus vulgaris*) or another large starch-storing seed transversely and then press the cut surface on a piece of two-sided tape attached to a microscope slide. Starch grains can be observed unstained (Fig. 29) or with polarizing microscopy (Fig. 30).

Compound starch grains are common in many plant organs including roots of sweet potato (*Ipomoea batatas*). Cut thin sections of a root of sweet potato, stain with I_2KI, and mount under a cover glass. Compound starch grains will stain purple (Fig. 31).

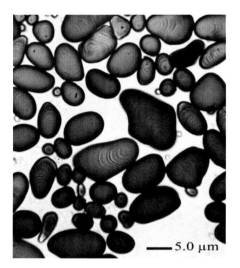

Figure 26. Starch grains from a potato (*Solanum tuberosum*) tuber stained with I_2KI.

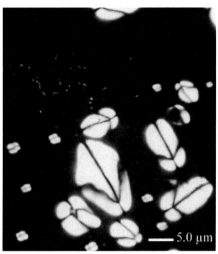

Figure 27. Unstained starch grains from a potato (*Solanum tuberosum*) tuber viewed with polarizing microscopy showing typical Maltese cross due to the crystalline nature of starch components.

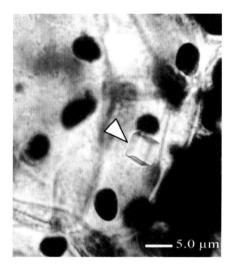

Figure 28. Starch grains and a cuboidal crystalloid of protein (arrowhead) within cells of a potato (*Solanum tuberosum*) tuber stained with I_2KI.

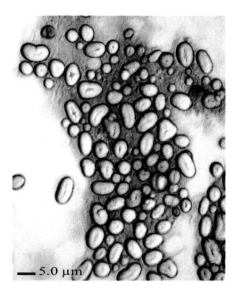

Figure 29. Unstained starch grains from an imbibed seed of bean (*Phaseolus vulgaris*).

Figure 30. Unstained starch grains from an imbibed seed of bean (*Phaseolus vulgaris*) viewed with polarizing microscopy.

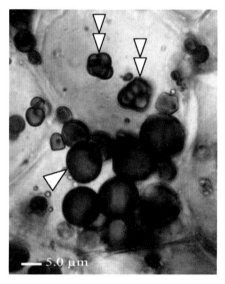

Figure 31. Section of sweet potato (*Ipomoea batatas*) root stained with I_2KI showing simple (arrowhead) and compound (double arrowheads) starch grains.

Box 3. Polarizing microscopy

Polarizing microscopy makes use of the property of crystals to rotate the direction of plane polarized light. Light can be considered to move in sinusoidal waves.

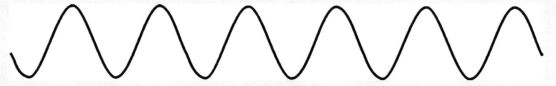

When seen from an end view, these waves occur at all 360 degrees.

The light source of the microscope emits light vibrating in all planes, as indicated by the various angles of the arrowheads emanating from the light bulb. A polarizing filter acts to screen out all light except the fraction that vibrates in one plane. This filter is inserted into the light path near the light source of a polarizing microscope. This can be as simple as setting one of the lenses from a pair of polaroid sunglasses over the light source at the base of the microscope. Now the light striking the specimen is plane polarized. A second filter (perhaps the other lens of the sunglasses) is placed in the light path beyond the specimen.

One of the filters can be rotated so that the second filter blocks off the light passed by the first one, a situation known as crossed polars, and a black field is obtained. A crystal will rotate the plane of the polarized light, allowing some of it to pass through the second filter. Thus, in a polarizing microscope, crystals appear as light bodies on a dark field. The increased contrast achieved by viewing an object against a dark field makes polarizing microscopy a very sensitive technique for detecting crystals. To be sure of seeing all the crystals present in the field of view, a rotating stage is necessary.

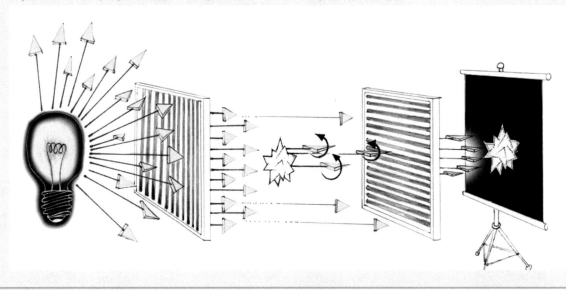

Other storage reserves

In addition to starch, plant cells may store **lipids** and **proteins**. Many seeds are often rich sources of one or more of these storage compounds.

Seeds of sunflower (*Helianthus annuus*) contain lipids and proteins that can be shown by staining thin sections. Obtain raw sunflower seeds and imbibe them in water for 24 hours. Remove seed coat and prepare thin sections of the **cotyledons** with a two-sided razor blade. Stain some for lipids and others for protein.

Lipid staining — Place sections in a Sudan stain (III or IV) (**see Appendix 2**) or a mixture of both on a microscope slide for five or more minutes. More stain will have to be added since it will evaporate quickly. Draw off the stain and mount sections in 50% glycerol under a cover glass. Lipids will stain orange–red (Figs. 32, 33).

An alternate source of material to illustrate lipids is the fruit of avocado (*Persea americana*). Remove the peel, prepare thin sections, and stain for lipids as described above for sunflower seeds. Parenchyma cells contain large numbers of lipid bodies that frequently are displaced from cells during cutting; these stain orange–red (Fig. 34).

Protein staining — Place sections of sunflower cotyledons in a drop of 1.0% amido black in 7.0% acetic acid (**see Appendix 2**) on a microscope slide for several minutes. Then rinse in 7.0% acetic acid and mount in water under a cover glass. Protein bodies stain blue (Figs. 35, 36).

Figure 32. Section of an imbibed seed of sunflower (*Helianthus annuus*) stained with Sudan III/IV showing numerous lipid bodies in cells of a cotyledon.

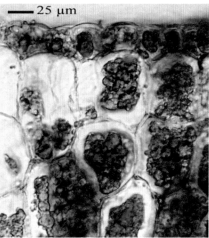

Figure 33. Section of an imbibed seed of sunflower (*Helianthus annuus*) stained with Sudan III/IV showing numerous lipid bodies in cells of a cotyledon at higher magnification.

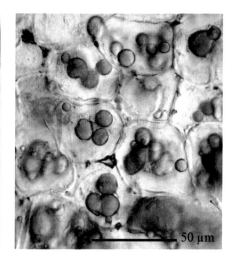

Figure 34. Section of an avocado (*Persea americana*) fruit stained with Sudan III/IV showing large lipid bodies.

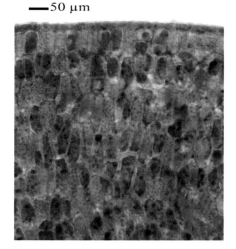

Figures 35. Section of an imbibed seed of sunflower (*Helianthus annuus*) stained with amido black showing many protein bodies in cotyledon cells.

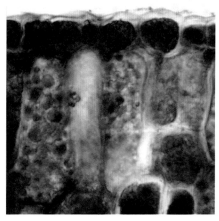

Figure 36. Section of an imbibed seed of sunflower (*Helianthus annuus*) stained with amido black showing many protein bodies in cotyledon cells.

An alternate method for protein staining is to place the section in 1.0% aqueous acid fuchsin on a microscope slide for several minutes. Rinse sections with 7.0% aqueous acetic acid and mount in water under a cover glass. Protein deposits should stain pink–red.

Tissue printing for protein

The distribution of proteins in large seeds such as those of some legumes can be demonstrated by tissue printing **(Box 4)**.

1. Imbibe seeds of *Phaseolus vulgaris* (common bean) for at least 24 hours.
2. Use forceps to place a piece of nitrocellulose membrane on top of several layers of Whatman filter paper in a Petri dish. Be careful not to touch the membrane with your fingers.
3. Cut an imbibed seed transversely towards one end and press the cut surface of the large piece firmly onto the nitrocellulose on different regions of the membrane.
4. Allow the membrane to dry briefly and then flood the surface with a 1.0% aqueous acid fuchsin solution. Let stand briefly and then flood the membrane several times with 7.0% acetic acid. Protein should stain pink–red (Fig. 37).

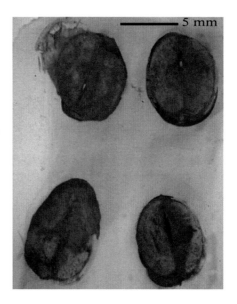

Figure 37. Tissue blot of imbibed seeds of bean (*Phaseolus vulgaris*) stained with acid fuchsin. The cotyledons show positive staining for proteins.

Box 4. Tissue printing

Reference: Reid, P.D., and Pont-Lezica, R.F. 1992. Tissue Printing. Academic Press, San Diego.

Hand sections of plant parts such as seeds, stems, etc. as well as whole organs such as leaves can be pressed onto a suitable substrate (e.g., nitrocellulose, chromatography paper, agarose gel) and then this substrate can be probed for various substances released from cut cells or surface structures such as trichomes (hairs). Probes can be histochemical stains, antibodies to proteins, substrates for enzymes, etc. This technique has been used effectively to localize proteins, nucleic acids, and various secretory compounds.

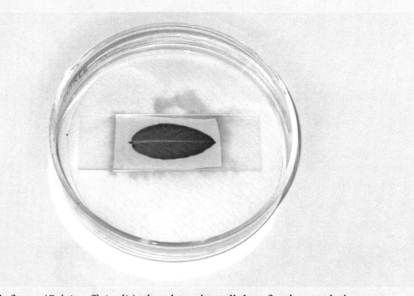

Leaf of sage (*Salvia officinalis*) placed on nitrocellulose for tissue printing

Ergastic substances

In addition to storage reserves, plant cells often synthesize and store substances with low or no metabolic turn-over. Some examples of these 'ergastic substances' are crystals, pigments, and phenolic compounds.

Calcium oxalate crystals

Calcium oxalate crystals occur in many plant species and may function to deter herbivores from eating plant parts. Crystals take many forms that are readily visible with light microscopy.

Prepare thin sections of *Tradescantia* spp. stems (at least 20 cm from the growing tip), place in a drop of water on a microscope slide and cover with a cover glass. **Prismatic** (Figs. 38, 39) and **raphide** (Fig. 39) crystals should be visible. The crystalline nature of these can be confirmed with polarizing microscopy.

Prepare thin sections of geranium (*Pelargonium hortorum*) stems or sweet potato (*Ipomoea batatas*) roots, mount them in water under a cover glass, and observe the **druse crystals** in the **cortex** (Fig. 40). Cleared leaves of geranium or other plants are also excellent to demonstrate druse crystals (Fig. 41). See **Boxes 21 and 22** for methods used to clear leaves.

Impatiens spp. petals have enlarged cells (**idioblasts**) that contain mucilage and a bundle of **raphide crystals**. To demonstrate this, cut pieces from petals and clear in 2.5% potassium hydroxide (KOH) by warming the slide until some of the pigments are released into the solution. The petal does not need to be completely clear. Mount in water under a cover glass. Compare the size of the idioblast cells with the surrounding cells and locate the bundle of raphides (Fig. 42).

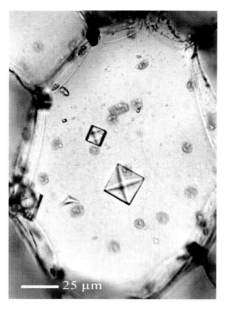

Figure 38. Section of *Tradescantia* sp. stem showing prismatic crystals.

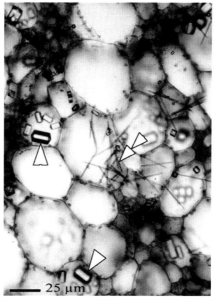

Figure 39. Section of *Tradescantia* sp. stem showing prismatic crystals (arrowheads) and raphide crystals (double arrowhead). Many of the raphide crystals have been displaced during sectioning.

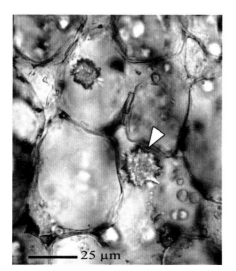

Figure 40. Section of sweet potato (*Ipomoea batatas*) root showing druse crystals (arrowhead).

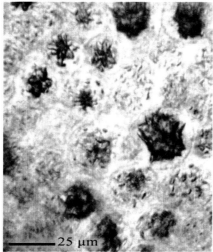

Figure 41. Cleared leaf of geranium (*Pelargonium hortorum*) showing druse crystals.

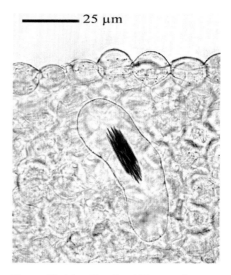

Figure 42. A bundle of raphide crystals within an enlarged cell (idioblast) of a cleared petal of *Impatiens* sp.

Calcium carbonate crystals

Rubber plant (*Ficus elastica*) leaves contain very unusual calcium carbonate crystals (**cystoliths**) in specialized large epidermal cells called **lithocysts**. The calcium carbonate is deposited around a cell wall extension (peg) that can be seen when the calcium carbonate is dissolved in vinegar or 10% acetic acid.

Prepare thin transverse sections of a portion of a rubber plant leaf and mount in water under a cover glass. Several lithocysts should be visible and these may have cystoliths of various sizes (Fig. 43). After locating cystoliths, draw some weak acetic acid under the cover glass and, after doing this several times, re-examine the sections. The peg with remnants of the cystolith and the outline of a **suberin** coat should be visible (Fig. 44).

Epidermal strips (peels) of leaves such as aluminum plant (*Pilea cadierei*) also show cystoliths in epidermal cells (Fig. 45). To prepare epidermal peels, fold the leaf over a finger, make a slit into the leaf with a sharp razor blade, and pull back the epidermis with fine forceps **(this is best illustrated in Box 19)**. Mount the peel in water on a microscope slide and add a cover glass. Cleared leaves **(see Boxes 21 and 22 for method)** of this species viewed with polarizing microscopy illustrate the number of such crystals that can develop (Fig. 46).

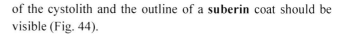

Figure 43. Transverse section of a rubber plant (*Ficus elastica*) leaf showing a large cystolith (arrowhead).

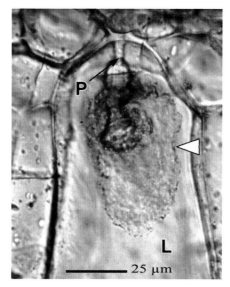

Figure 44. Transverse section of a rubber plant (*Ficus elastica*) leaf showing the suberin coat that had surrounded the crystal. (arrowhead) after treatment with acetic acid. The peg or wall ingrowth (P) as well as the enlarged cell, the lithocyst (L), are evident.

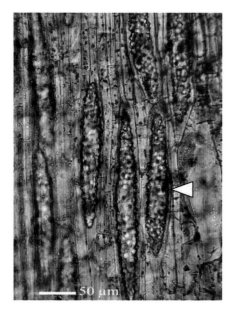

Figure 45. Epidermal strip of a aluminum plant (*Pilea cadierei*) leaf showing large cystolith crystals (arrowhead). Photo courtesy of Britta Ng.

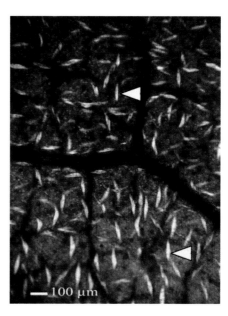

Figure 46. Cleared leaf of aluminum plant (*Pilea cadierei*) viewed with polarizing microscopy showing numerous cystolith crystals (arrowheads). Photo courtesy of Aidon Pyne.

Vacuolar pigments

Anthocyanins are red or blue water-soluble pigments that are found within vacuoles. Roots of garden beet (*Beta vulgaris*), some fruits, and many red and purple petals have cells with anthocyanin pigments in their vacuoles.

Prepare thin sections of a garden beet root, mount in water under a cover glass, and observe parenchyma cells containing anthocyanins (Fig. 47). Many cells will appear clear because the pigments have been released from the vacuoles of cells damaged during sectioning.

Cells in the enlarged receptacle (fleshy portion) of strawberry (*Fragaria ananassa*) aggregate fruits contain anthocyanin pigments that give the red colour when ripened. Section a portion of the receptacle, mount in water under a cover glass, and observe cells with anthocyanin pigments (Fig. 48).

Red and purple flower petals of species such as geranium (*Pelargonium hortorum*) and African violet (*Saintpaulia ionantha*) are excellent to show anthocyanins within vacuoles. Tear petals or section them using Styrofoam as a support; mount specimens in water under a cover glass. Examine the edge of the torn petals or sections in which the anthocyanin pigment has remained within cells (Figs. 49, 50).

Tannins; polyphenols

These compounds are abundant in some plant species. They may be located within vacuoles, the cell cytoplasm, and the cell wall. Tannins are glycosides containing polyhydroxyphenols or their derivatives. They are colourless and have to be stained in order to be detected. Anhydrous forms are referred to as phlobaphenes; these have natural yellow, red, or brown colours and can be identified in unstained sections. Polyphenols are aromatic compounds, many of which will fluoresce when viewed with UV light. A number of staining methods are also used to detect these in plant cells. All of these compounds play many roles in plants, among which deterring predators and pathogenic organisms are important. Some act as antioxidants, scavenging reactive oxygen species that are harmful to plant cells.

Figure 47. Unstained section of beet (*Beta vulgaris*) root showing cells with vacuoles filled with anthocyanin pigments. Clear cells are those in which the pigments have escaped from the vacuoles during sectioning.

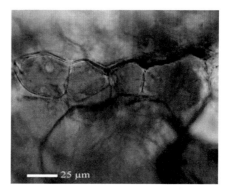

Figure 48. Unstained section of the receptacle of a strawberry (*Fragaria ananassa*) fruit showing anthocyanin pigments.

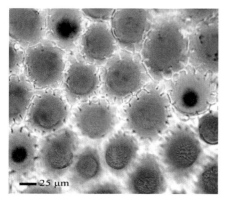

Figure 49. Geranium (*Pelargonium hortorum*) petal showing vacuolar anthocyanins.

Figure 50. African violet (*Saintpaulia ionantha*) petal showing cells with anthocyanin pigments.

Prepare transverse sections of Dutchman's pipe (*Aristolochia durior*) stems and mount in water under a cover glass. Many cells scattered throughout the stem contain orange deposits of phlobaphenes (Fig. 51).

Coleus (*Coleus blumei*) stems and leaves have large amounts of the phenolic compound rosmarinic acid in many parenchyma cells. Prepare transverse sections of a young stem and mount in water under a cover glass. View with UV light using an epifluorescence microscope. Cells in the epidermis and cortex contain rosmarinic acid in their vacuoles (Fig. 52).

Leaves and petioles of oak (*Quercus* spp.) and maple (*Acer* spp.) contain tannins that are colourless but these can be readily stained with vanillin-HCl **(see Appendix 2)** that binds these compounds and renders them pink–red in colour. Prepare transverse sections of petioles of these tree species, mount some in water under a cover glass, and stain others in vanillin-HCl **(see Appendix 2)**. Mount stained sections in a drop of the stain under a cover glass. Compare the unstained and stained petiole sections of oak (Figs. 53, 54) and maple (Figs. 55, 56).

The needles of all pine (*Pinus*) species contain large amounts of tannins. Section needles transversely, mount some in water under a cover glass, and stain others with vanillin-HCl. Compare unstained (Fig. 57) with stained (Fig. 58) sections.

Roots of bananas (*Musa* spp.), like roots of many species, contain tannins. Prepare transverse sections, mount some in water under a cover glass, and stain others with vanillin-HCl. Compare unstained (Fig. 59) and stained (Fig. 60) sections.

Figure 51. Section of Dutchman's pipe (*Aristolochia durior*) stem that has been lightly stained with Toluidine Blue O. Naturally orange deposits of phlobaphenes (arrowheads) are present.

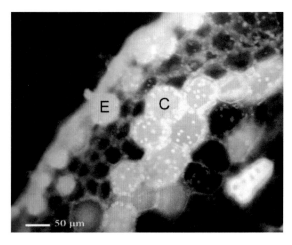

Figure 52. Section of coleus (*Coleus blumei*) stem mounted in water and viewed with UV light using an epifluorescence microscope. Cells in the epidermis (E) and cortex (C) contain rosmarinic acid that fluoresces in their vacuoles.

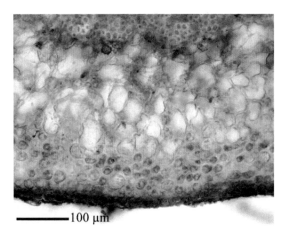

Figure 53. Unstained transverse section of oak (*Quercus* sp.) petiole.

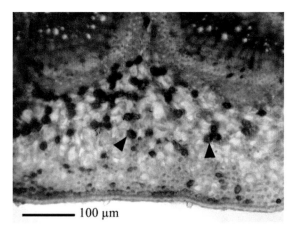

Figure 54. Transverse section of oak (*Quercus* sp.) petiole stained with vanillin-HCl showing distribution of cells containing tannins (arrowheads).

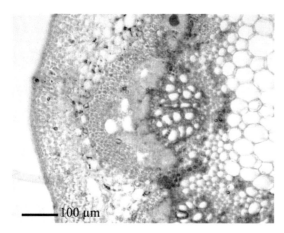

Figure 55. Unstained transverse section of maple (*Acer* sp.) petiole.

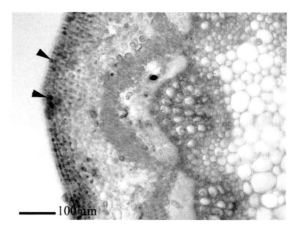

Figure 56. Transverse section of maple (*Acer* sp.) petiole stained with vanillin-HCl showing cells containing tannins (arrowheads).

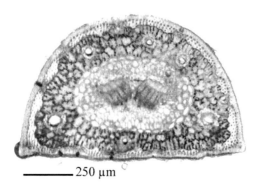

Figure 57. Unstained transverse section of pine (*Pinus sylvestris*) needle.

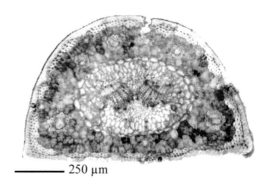

Figure 58. Transverse section of pine (*Pinus sylvestris*) needle stained with vanillin-HCl. Tannins are found throughout tissues of the needle.

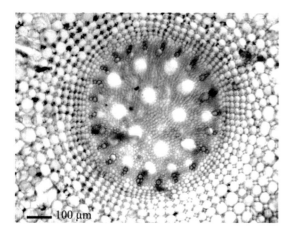

Figure 59. Unstained transverse section of banana (*Musa* sp.) root.

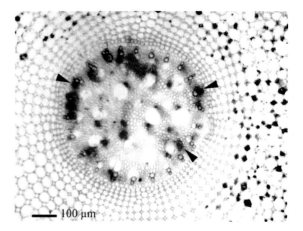

Figure 60. Transverse section of banana (*Musa* sp.) root stained with vanillin-HCl. Tannin-containing cells (arrowheads) are evident.

Chapter 4
Cell types in simple tissues

There are various ways of observing cell types and characterizing their structure in plant organs. Hand sectioning followed by staining with Toluidine Blue O is one method to determine the nature of the walls of plant cells and, therefore, the cell types present. This stain **(see Appendix 2)** will be used throughout the remainder of the book. **Box 5** gives some details concerning Toluidine Blue O and outlines the procedure for its use.

Parenchyma

Parenchyma cells are the basic cell type in plants and are the type from which other cell types develop during differentiation processes. Some details of this cell type are summarized in **Box 6**. Parenchyma cells have a thin primary cell wall, a nucleus, various amounts of cytoplasm, and one or more vacuoles. Usually intercellular spaces occur in the tissue.

Stems, petioles, or roots of a variety of plants can be used to demonstrate parenchyma cells. Cut thin sections of the petiole of celery (*Apium graveolens*) or the stem of any plant available. Float the sections in a small dish of water and select the thinnest sections. Mount some of these in water under a cover glass and stain others with Toluidine Blue O using the method outlined in **Box 5**. Compare the appearance of the parenchyma cells in the pith or cortex in unstained and stained sections. The primary cell walls of parenchyma cells should stain a pink–purple colour with Toluidine Blue O because of the abundance of pectic substances (Fig. 61).

Parenchyma cells in some aquatic plants and plants growing in moist environments often develop shapes that result in the formation of large air spaces (lacunae) between adjacent cells. The resulting tissue is a specialized type of parenchyma called **aerenchyma**, shown in **midribs** and petioles of canna lily (*Canna generalis*). Cut sections of either the midrib or petiole (only partial sections are needed), mount in water under a cover glass, and observe the branched parenchyma cells (Fig. 62). These cells are living and contain plastids and other organelles (Fig. 63).

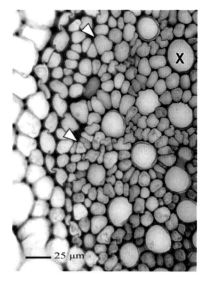

Figure 61. Transverse section of celery (*Apium graveolens*) petiole stained with TBO showing parenchyma cells (arrowheads) and lignified tracheary elements in the xylem (X).

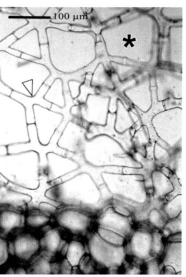

Figure 62. Unstained transverse section of the petiole of a canna lily (*Canna generalis*) leaf. Branched parenchyma cells (arrowhead) are separated by large air lacunae (∗).

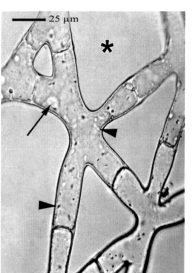

Figure 63. Branched parenchyma cells separated by large air lacunae (∗) in a petiole of a canna lily (*Canna generalis*) leaf viewed at higher magnification. A nucleus (arrow) and small chloroplasts (arrowheads) are evident.

Box 5. Staining hand sections with Toluidine Blue O (TBO)

Toluidine Blue O is a polychromatic dye (i.e., different components of the cell wall stain different colours). Walls containing **pectic substances** but not **lignin** stain a pink–purple colour, whereas walls containing lignin stain various shades of blue or blue–green. The simple and rapid staining process of fresh hand sections using Toluidine Blue O produces results that are usually very colourful and encouraging to students who often have observed only standard prepared microscope slides in biology classes. This method of staining plant tissues was introduced in the following publication:

O'Brien, T.P., Feder, N., and McCully, M.E. 1964. Polychromatic staining of plant cell walls by Toluidine Blue O. Protoplasma 59: 368–373.

Procedure (refer also to Chapter 2 and Appendix 2)

1. Cut sections with a two-sided razor blade that has been snapped into two pieces prior to removal from paper packaging, and float them off the blade into a small dish of water.
2. Pick up several thin sections (nearly colourless) with a small paintbrush and place either in a small dish with stain or on a slide in a drop of stain. The time for staining varies with the tissue but generally around one minute is sufficient.
3. If sections have been stained in a dish, pick up section with a brush and place in a drop of water on a microscope slide. After one minute, draw off water with a Kimwipe or filter paper and add another drop of water and a cover glass. View specimen with white light.
4. If sections have been stained on a slide, draw off the stain, replace with water, and draw this off with Kimwipe. Add a drop of water and mount under a cover glass. View specimen with white light.

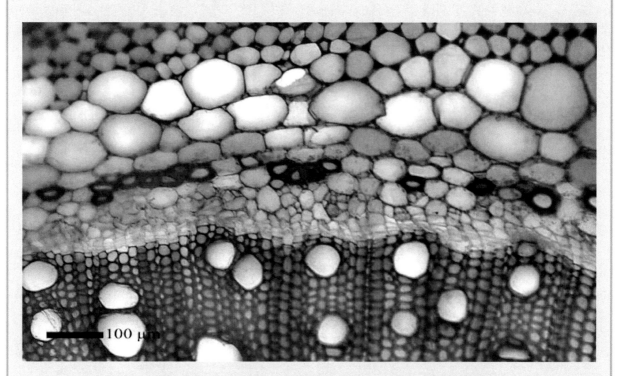

Cross section of a Jimson weed (*Datura stramonium*) stem stained with TBO.

Box 6. Parenchyma

Parenchyma cells are located in all plant organs and perform various functions such as photosynthesis, storage, secretion, and short distance transport. They assume various shapes depending on their position in organs and their function. Parenchyma cells are **totipotent** (i.e., they have a living protoplast and are capable of regenerating parts of the plant or the entire plant under suitable conditions). This fact has contributed to the development of *in vitro* propagation of many plant species. Most parenchyma cells have the following features:

- living
- thin primary cell wall that is usually unlignified

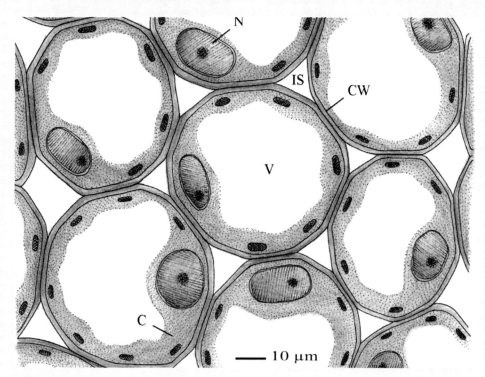

Parenchyma cells have a thin primary cell wall (CW) that is even thinner than illustrated, a nucleus (N), various amounts of cytoplasm (C), and one or more vacuoles (V). Usually intercellular spaces (IS) occur in the tissue.

Collenchyma

Petioles of celery (*Apium graveolens*) have large patches of **angular collenchyma** towards the periphery of this elongated organ. These are the 'strings' that get stuck between your teeth when eating celery. Trim the petiole **(Chapter 2)**, leaving a 2 mm × 2 mm face including the collenchyma. Cut thin sections; leave some unstained and stain others with Toluidine Blue O (**see Box 5**), and locate the patches of collenchyma. In unstained sections, collenchyma cell walls appear white because they are highly hydrated due to the presence of pectic substances (Fig. 64). The angular collenchyma cell walls stain a pink–purple when stained with Toluidine Blue O (Fig. 65). At higher magnification, the angular nature of the thickened cell walls is evident (Fig. 66). Soybean (*Glycine max*) stems also have angular collenchyma (Fig. 67).

Stems and petioles of jimson weed (*Datura stramonium*) and *Alstroemeria* spp. are excellent for demonstrating the three types of collenchyma (**Box 7**). Cut thin sections of either stems or petioles, mount some under a cover glass in water, and stain others with Toluidine Blue O. The epidermis in *Alstroemeria* spp. consists of **lamellar collenchyma** (Fig. 68). Both **angular** and **lacunar collenchyma** are present in the peripheral portion of the cortex of Jimson weed stems (Fig. 69).

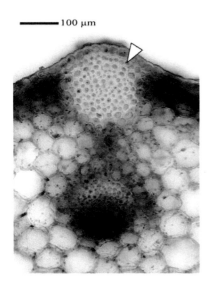

Figure 64. Unstained transverse section of celery (*Apium graveolens*) petiole showing a bundle of angular collenchyma (arrowhead).

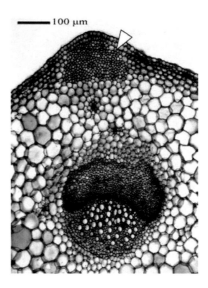

Figure 65. Section of celery (*Apium graveolens*) petiole stained with TBO showing a bundle of angular collenchyma (arrowhead).

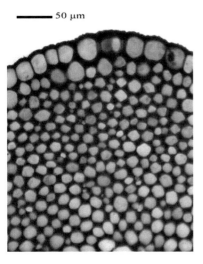

Figure 66. High magnification of angular collenchyma from a celery (*Apium graveolens*) petiole stained with TBO, indicating an unusually large amount of pectic substances in the cell walls.

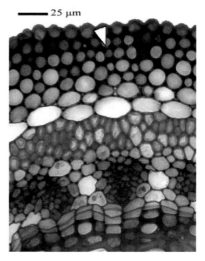

Figure 67. Transverse section of soybean (*Glycine max*) stem stained with TBO showing angular collenchyma (arrowhead).

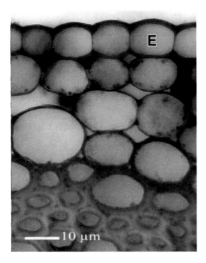

Figure 68. Transverse section of *Alstroemeria* sp. petiole stained with TBO showing lamellar collenchyma in the epidermis (E).

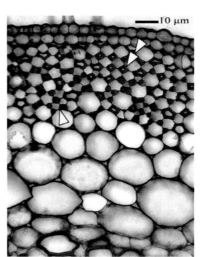

Figure 69. Transverse section of Jimson weed (*Datura stramonium*) stem stained with TBO showing angular (arrowhead) and lacunar (double arrowhead) collenchyma.

Box 7. Collenchyma

Collenchyma cells differentiate from parenchyma cells and are characterized by being alive at maturity and having unevenly thickened primary cell walls. The pattern in which the thickened primary cell wall is deposited is the basis for classifying collenchyma as **angular**, **lacunar,** or **lamellar**. All types may be present in the same organ. Angular collenchyma cells (AN) have wall thickenings in their corners; lacunar collenchyma (LC) is similar except that intercellular spaces (IS) are present between cells. Lamellar collenchyma (LM) cells have thickenings on the inner and outer tangential cell walls. Collenchyma cells are located at the periphery of some aerial organs (stems, petioles, flower stalks) and strengthen these organs while they are still elongating. Collenchyma cell walls are usually not lignified but may become so as the organ ages.

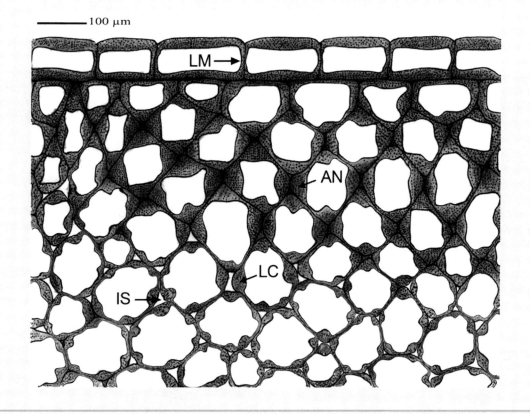

Sclerenchyma

Sclerenchyma is a simple tissue consisting of either fibres or sclereids **(see Box 8)**.

There are a number of ways of preparing plant material to observe fibres and sclereids: staining sections of fresh material with either Toluidine Blue O or phloroglucinol-HCl **(see Appendix 2)**, macerating material, and/or clearing entire plant parts.

Fibres

Leaves of the snake plant (*Sansevieria zeylanica*) contain bundles of fibres that provide strength to these elongated organs. Prepare thin transverse sections of a portion of a leaf, stain in Toluidine Blue O, and observe the bundles of fibres located in the mesophyll (Fig. 70) and the fibres associated with the phloem of the vascular bundles (Fig. 71). The thickened secondary cell walls stain blue or blue–green because of the presence of lignin. Careful observations at high magnification will reveal the narrow pits **(see Box 8)** in these walls.

Prepare thin longitudinal sections of a portion of a leaf and stain in Toluidine Blue O. The narrow, elongated shape of the fibres should be evident (Fig. 72).

Macerate sections of snake plant leaves as outlined in **Box 9**. Tease apart the white strands in the macerated material, stain in Toluidine Blue O, and determine the length and shape of the fibres in these strands. When viewed with a polarizing microscope, the walls of the fibre give evidence of a large quantity of the crystalline form of cellulose (Fig. 73).

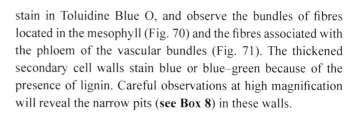

Figure 70. Bundle of fibres (arrowhead) within the mesophyll of a snake plant (*Sansevieria zeylanica*) leaf. Transverse section stained with TBO.

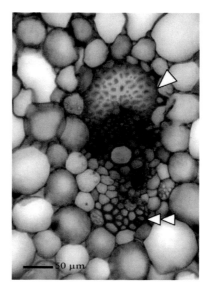

Figure 71. Bundle cap of fibres (arrowhead) in the phloem of a snake plant (*Sansevieria zeylanica*) leaf. Thinner walled fibres (double arrowheads) are also associated with the primary xylem. Transverse section stained with TBO.

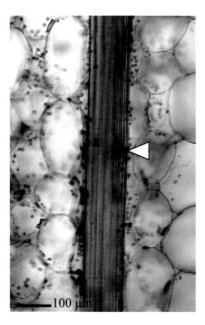

Figure 72. Longitudinal section of a snake plant (*Sansevieria zeylanica*) leaf stained with TBO showing a group of fibres (arrowhead).

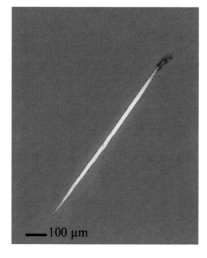

Figure 73. Fibre from a macerated leaf of snake plant (*Sansevieria zeylanica*) viewed with polarizing microscopy.

Box 8. Sclerenchyma

Sclerenchyma cells can be found in most plant organs but are usually absent in roots. These cells develop a secondary cell wall (SW) which, along with the primary cell wall (PW), usually becomes lignified. Pits (P) are present in the secondary cell wall. The physical and chemical characteristics of these walls contribute to their function as support cells. The cells are normally dead at maturity, and can have a variety of shapes. The two major types of sclerenchyma cells, **fibres** and **sclereids**, differ primarily in their shape. Fibres are usually very elongated, narrow cells, whereas sclereids are often shorter and can have a variety of shapes. Several types of sclereids are important components of seed coats, providing a barrier to the ingress of microorganisms.

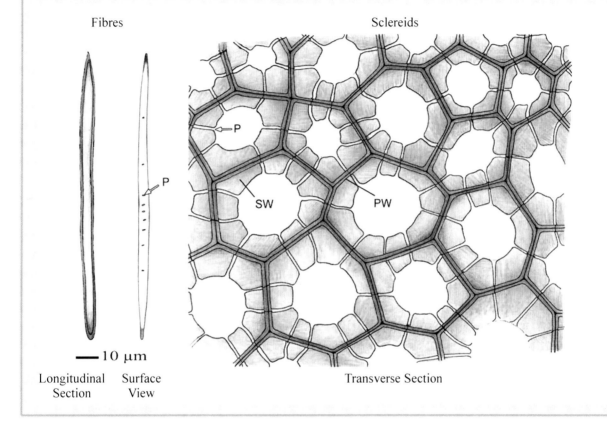

Sclereids

Fruits of pear (*Pyrus communis*) and avocado (*Persea americana*) contain many **brachysclereids** (stone cells) that give them their gritty texture. Prepare thin sections of the flesh of a pear fruit or, if the fruit is very ripe, simply remove some of the flesh with a scalpel. Stain some sections with Toluidine Blue O and others with phloroglucinol-HCl to confirm that these cells have lignified walls. Clusters of brachysclereids should be evident. These will have very thick, lignified secondary cell walls with simple pits, many of which have branches; the term **ramiform pits** is often used to denote the latter (Fig. 74). Sections of avocado fruits including the peel also show very thick-walled brachysclereids, the walls of which stain blue with Toluidine Blue O (Fig. 75) and red with phloroglucinol-HCl (Fig. 76), again showing the lignified nature of the thick secondary cell wall.

Prepare thin sections of the stem of a wax plant (*Hoya carnosa*); stain with Toluidine Blue O or phloroglucinol-HCl. Parenchyma cells in the cortex and pith undergo differentiation into brachysclereids (Fig. 77) and, therefore, depending on the internode of the stem sectioned, stages in the formation of these cells can be observed. At higher magnification details of cortical brachysclereids (Fig. 78) and brachysclereids in the pith (Fig. 79) can be seen.

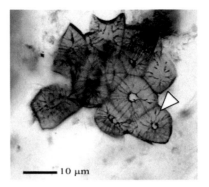

Figure 74. Section of pear (*Pyrus communis*) fruit stained with phloroglucinol-HCl. Thick-walled, lignified brachysclereids with pit canals (arrowhead) are evident. Photo courtesy of Aidon Pyne.

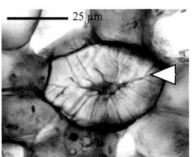

Figure 75. Section of avocado (*Persea americana*) fruit near the peel stained with TBO. Pit canals (arrowhead), some of which are branched, are evident in the thick, lignified cell wall of the brachysclereid.

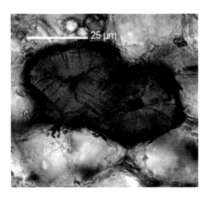

Figure 76. Section of avocado (*Persea americana*) fruit near the peel stained with phloroglucinol-HCl showing two brachsclereids with thick, lignified cell walls.

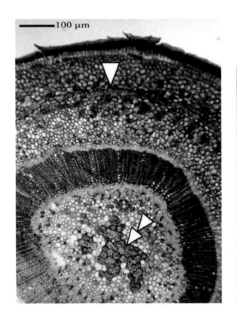

Figure 77. Transverse section of wax plant (*Hoya carnosa*) stem stained with TBO. Brachysclereids are present in the cortex (arrowhead) and the pith (double arrowheads).

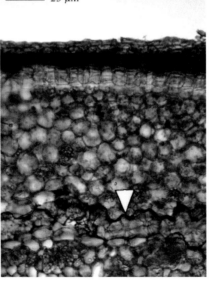

Figure 78. Transverse section of wax plant (*Hoya carnosa*) stem stained with TBO showing brachysclereids (arrowhead) in the cortex.

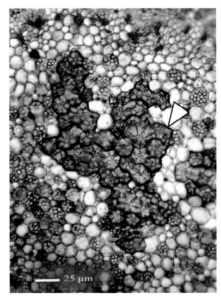

Figure 79. Transverse section of wax plant (*Hoya carnosa*) stem stained with TBO showing brachysclereids (arrowhead) in the pith.

Box 9. Maceration technique

Prepare macerating fluid as follows:

1 part 30% solution of hydrogen peroxide (H_2O_2)

4 parts water

5 parts glacial acetic acid

(This solution should be prepared just before use and care must be taken in handling it because it is caustic.)

A. **Soft tissues:** Cut samples into small pieces, place in vials that can be capped, and add macerating fluid. **Caps should be applied loosely, since pressure will build up during heating.** Place vials in an oven at 55–60 °C for 12–24 hours. Test a portion of the material to see if it can be teased apart easily. If not, add fresh macerating fluid and place back in the oven. When maceration is complete, remove macerating fluid and gently wash material several times with water. Material can be stored in 70% ethanol for months.

B. **Seed coats:** Imbibe seeds in water so that the seed coat can be removed, cut the coat into pieces, and then place them in macerating fluid and follow the protocol above.

C. **Woody tissues:** Remove bark from woody twigs and cut the xylem (wood) into toothpick sized pieces. Place in macerating fluid following the protocol above. It may take several days for the maceration to be completed and one or more changes of macerating fluid may be needed. When maceration is complete, wood samples will be whitish to translucent but should remain intact unless teased apart with probes. Wash samples several times with water, and with a small volume of water, vigorously shake the vials to disperse cells. Store in 70% ethanol.

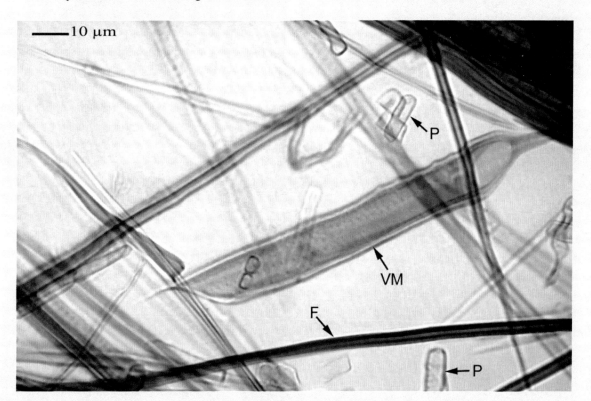

Macerated wood of oak (*Quercus* sp.) consisting of lignified parenchyma cells (P), a vessel member (VM), and fibres (F).

Macerated **seed coats (testas)** of legume species **(for method see Box 9)** are excellent to show various types of sclereids. Place a sample of macerated seed coat of garden pea (*Pisum sativum*) on a microscope slide, stain with Toluidine Blue O, and observe the **osteosclereids**, bone-shaped cells, sometimes called hour-glass cells (Fig. 80) and **macrosclereids**, elongated rod-like cells (Fig. 81). Osteosclereids and macrosclereids show birefringence from crystallized cellulose when viewed with polarizing microscopy (Figs. 80, 81).

The thin seed coat of peanut (*Arachis hypogaea*) consists of very interesting strengthening cells. Although they resemble brachysclereids, the thickened walls are not lignified and, therefore, they are modified parenchyma cells. Choose portions of seed coat from the maceration in which the cells are still attached. Place these on a microscope slide and stain with Toluidine Blue O. The unevenly thickened secondary cells walls stain the distinctive pink–purple colour indicative of pectic substances (Fig. 82).

Leaves of split-leaf philodendron (*Monstera deliciosa*) that have been partially macerated for 12–24 hours **(see Box 9)** show branched sclereids with thin, elongated arms (**trichosclereids**). Mount pieces of macerated leaf in water under a cover glass and observe the trichosclereids (Figs. 83, 84); compare the size of these to the surrounding cells. If leaves of *Monstera deliciosa* are not available, water lilies such as *Nymphaea odorata* can be used. The **astrosclereids** in this species are particularly interesting because they have small prismatic crystals of calcium oxalate embedded in their cell walls (Fig. 85).

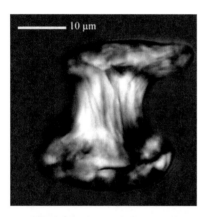

Figure 80. Osteosclereid from macerated seed coat of pea (*Pisum sativum*) viewed with polarizing microscopy. The thick cell walls show birefringence.

Figure 81. Macrosclereids from macerated seed coat of pea (*Pisum sativum*) viewed with polarizing microscopy. Cell walls show birefringence.

Figure 82. Macerated seed coat of peanut (*Arachis hypogaea*) stained with TBO showing parenchyma cells with modified walls.

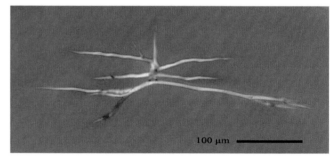

Figure 83. Trichosclereid from macerated leaf of split-leaf philodendron (*Monstera deliciosa*) viewed with polarizing microscsopy.

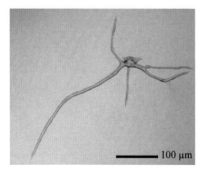

Figure 84. Trichosclereid from macerated leaf of split-leaf philodendron (*Monstera deliciosa*) viewed with polarizing microscopy.

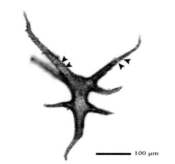

Figure 85. Astrosclereid from macerated leaf of water lily (*Nymphaea odorata*) stained with TBO. Calcium oxalate crystals (arrowheads) are evident.

Chapter 5
Complex Tissues

Xylem

Xylem, the transport system for water and mineral ions in plants, is a complex tissue consisting of several cell types **(Box 10)**. Stems of a variety of plants can be used to demonstrate the characteristics of xylem tissue. Sunflower (*Helianthus annuus*) and soybean (*Glycine max*) stems are particularly good to show components of both primary and secondary xylem.

Cut transverse sections of stems two to three internodes below the shoot apex and stain in Toluidine Blue O. Find the vascular bundles with the xylem located towards the pith (centre of the stem). The walls of tracheary elements and fibres stain blue to blue–green because of the presence of lignin, whereas the walls of xylem parenchyma cells stain pink–purple because of a greater proportion of pectic substances and lack of lignin (Figs. 86, 87). The xylem parenchyma cells remain alive and contain organelles such as chloroplasts. Sections of sunflower stems taken three to four internodes below the shoot apex illustrate the maturation pattern of the primary xylem and the arrangement of **protoxylem** and **metaxylem** in most stems (Figs. 88, 89).

Figure 90 illustrates the differentiation of protoxylem and metaxylem with reference to the shoot apex. The first functional xylem **tracheary elements** (protoxylem) are located towards the centre of the stem and have either **annular** or **helical** (**spiral**) secondary cell wall thickenings. Later primary xylem (metaxylem) consists of tracheary elements with **scalariform, reticulate,** and **pitted** secondary cell wall patterns.

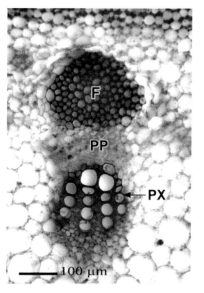

Figure 87. Transverse section of sunflower (*Helianthus annuus*) stem stained with TBO showing a vascular bundle with primary xylem (PX) and primary phloem (PP). A cap of fibres (F) is part of the primary phloem.

Figure 88. Transverse section of sunflower (*Helianthus annuus*) stem stained with TBO taken 1–3 cm from the shoot apex. Mature protoxylem (arrowhead) and some immature metaxylem elements lacking a lignified secondary cell wall (double arrowhead) are present.

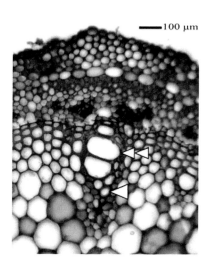

Figure 86. Transverse section of soybean (*Glycine max*) stem stained with TBO. Protoxylem (arrowhead) and metaxylem (double arrowhead) are evident in the primary xylem.

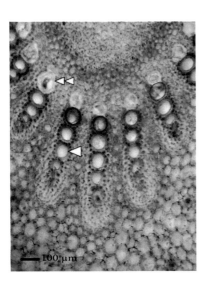

Figure 89. Transverse section of sunflower (*Helianthus annuus*) stem taken 1–3 cm from the shoot apex stained with phloroglucinol-HCl. The protoxylem (arrowhead) is mature. Some metaxylem elements have formed a thick wall, but have not yet deposited lignin (double arrowhead).

Cut longitudinal sections **(see Chapter 2)** of similar stem internodes used for the transverse sections, stain either with Toluidine Blue O or phloroglucinol - HCl, and examine for the presence of tracheary elements. The secondary cell walls of these elements are deposited in varying patterns depending on the time at which the elements were formed (refer to Fig. 90). Tracheary elements in the **protoxylem** have either helical (spiral) or annular (Figs. 91, 92) secondary wall patterns; these configurations allow these elements to stretch during the elongation of internodes. Tracheary elements in the **metaxylem** have scalariform (Fig. 93), reticulate, or pitted (Fig. 94) secondary cell wall patterns; these walls were deposited in tracheary elements of internodes that had ceased elongation. An informative, simple exercise to illustrate the temporal development of secondary cell wall patterns in tracheary elements is outlined in **Box 11**.

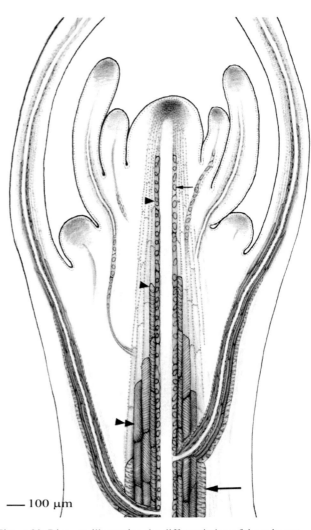

Figure 90. Diagram illustrating the differentiation of the primary xylem in a dicotyledonous stem as viewed in a longitudinal section. The first xylem to become functional is the protoxylem consisting of tracheary members (elements) with either annular (small arrow) or helical (arrowheads) secondary cell wall thickenings. Later, metaxylem consisting of tracheary members (elements) with scalariform (double arrowheads) or pitted (large arrow) secondary cell wall patterns mature.

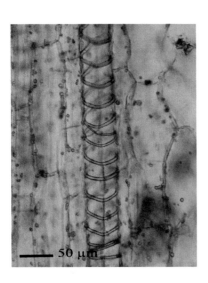

Figure 91. Unstained longitudinal section of coleus (*Coleus blumei*) stem showing helical (spiral) secondary cell wall deposition in a tracheary element.

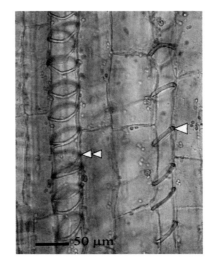

Figure 92. Unstained longitudinal section of coleus (*Coleus blumei*) stem showing annular secondary cell wall deposition in a tracheary element (arrowhead). One tracheary element (double arrowhead) has both annular and helical secondary cell wall depositions.

Complex Tissues

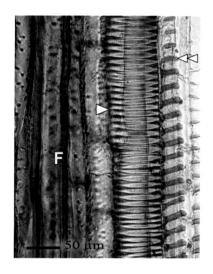

Figure 93. Longitudinal section of jimson weed (*Datura stramonium*) stem stained with phloroglucinol-HCl showing a wide-diameter tracheary element with scalariform secondary cell wall deposition (arrowhead), and two narrow-diameter elements with annular secondary cell wall deposition (double arrowhead). Thick-walled fibres (F) are also present.

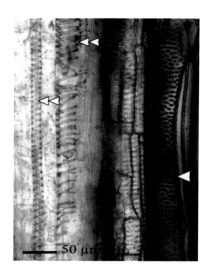

Figure 94. Longitudinal section of jimson weed (*Datura stramonium*) stem stained with phloroglucinol-HCl showing several tracheary elements with pitted secondary cell walls (arrowhead) and two with helical secondary cell walls (double arrowheads).

Box 10. Xylem

All vascular plants develop **primary xylem** during organ development. The component cells of the primary xylem system develop from part of the primary meristem, **procambium** (sometimes referred to as **provascular tissue**). The primary xylem is composed of **protoxylem** and **metaxylem**. Plants that undergo secondary growth form **secondary xylem** that is initiated by the **vascular cambium**.

The conducting cells of both primary and secondary xylem are the **tracheary elements** of which two types (**tracheids** and **vessel members [elements]**) may develop depending on the plant group. Most gymnosperms lack vessel members. All tracheary elements develop a thick lignified secondary cell wall and are dead at maturity. Vessel members develop openings in their end walls that lack both primary and secondary cell walls. These openings, **perforation plates**, can be of several types. Two or more vessel members join end to end to form a **vessel**.

Other cell types present in both primary and secondary xylem are parenchyma cells and often **fibres**.

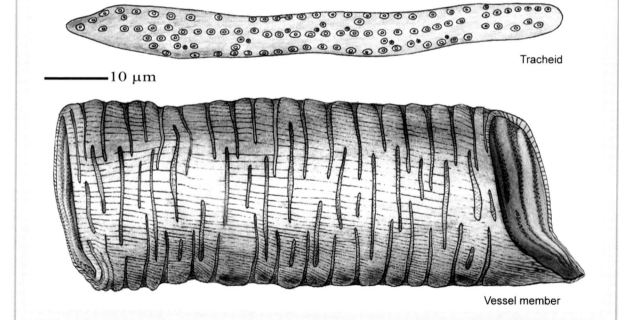

43

Box 11. Comparison of tracheary elements of elongating and non-elongating sunflower stem internodes

Sunflower (*Helianthus annuus*) plants that have reached several internodes in length should be selected and the stems should be marked with a permanent ink-based felt marker at 5 mm intervals for a distance of 10 cm from the shoot apex. Allow the plants to grow for at least one week and then measure the distances between the marks to determine the present regions of elongation and non-elongation along the stem. From these measurements, the region of the stem currently in the process of elongation can be predicted.

Excise one segment 1 cm in length from the region of the stem predicted to be in rapid growth and a 1 cm segment from a non-elongating region, cut them in half lengthwise, and place them in macerating fluid following the instructions for soft tissues outlined in **Box 9**. After maceration has been completed, place a drop of macerated material on a slide and add a small drop of Toluidine Blue O. Do not rinse. Compare the two samples in order to identifty types of secondary wall patterns of tracheary elements.

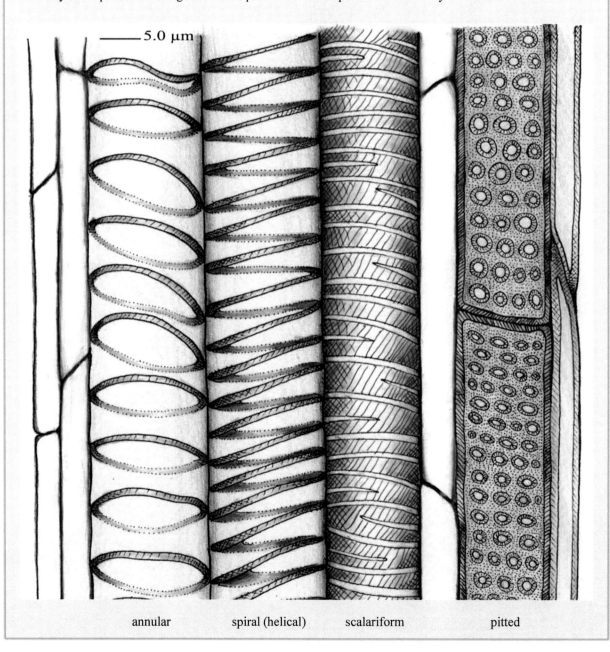

annular   spiral (helical)   scalariform   pitted

Macerated **wood** is excellent material to illustrate the complex nature of xylem, and particularly the detailed structural features of tracheary elements. Wood of any species can be macerated as outlined in **Box 9** and various exercises can be designed with this material. To illustrate some of the basic features of wood, species should be chosen that show variation in cell types. Examples are listed below.

Pine (*Pinus* sp.) — Xylem of pine species is rather simple in structure, consisting mainly of tracheids with large simple and **bordered pits**. Vessel members are absent. After making a maceration, it is sometimes necessary to stain the cells with TBO to improve their contrast. This can be done by mounting the sections in dilute TBO (about one-quarter strength) or by adding a very small droplet of the stain to a water mount and mixing them together prior to adding the cover glass. Because the cells concentrate the stain from the surrounding solution, they will show up well against the faint blue background without the necessity of drawing off the stain and rinsing the material (which removes the cells as well as the dye). Observe the elongated tracheids with many bordered pits (Fig. 95). A few parenchyma cells with lignified secondary cell walls and fibres may be present.

Tulip tree (*Liriodendron tulipifera*) — Xylem of this species consists of both tracheids and vessel members as conducting elements, as well as fibres and lignified parenchyma cells. The vessel members have **scalariform perforation plates** (Fig. 96).

Oak (*Quercus* sp.) — Both tracheids and short vessel members are present, the latter with **simple perforation plates** (Fig. 97).

Grape (*Vitis vinifera*) — A feature of grape secondary xylem is the development of **septate fibres** (Fig. 98).

Ephedra (*Ephedra* sp.) — This specialized gymnosperm has both tracheids and vessel members, the latter with a unique perforation plate (**ephedroid or foraminate**) consisting of a series of open holes (Fig. 99).

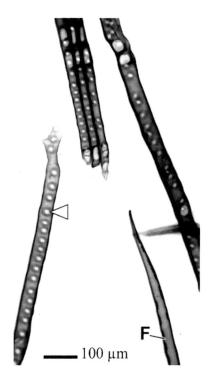

Figure 95. Tracheids from macerated xylem of white pine (*Pinus strobus*) showing the large bordered pits (arrowhead). A portion of a fibre (F) is also present. Stained with TBO.

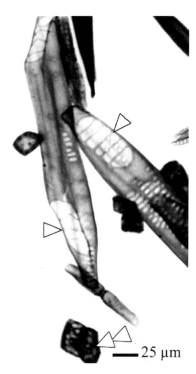

Figure 96. Vessel members from macerated xylem of the tulip tree (*Liriodendron tulipifera*) showing scalariform perforation plates (arrowheads). Small parenchyma cells with lignified secondary cell walls are also present (double arrowhead). Stained with TBO.

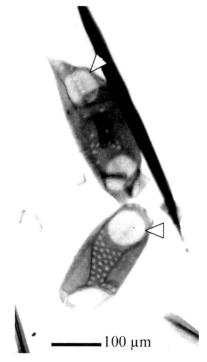

Figure 97. Vessel members from macerated xylem of oak (*Quercus* sp.) showing simple peforation plates (arrowheads). Stained with TBO.

To stimulate interest, students can be encouraged to select their own species to examine and keys can be developed by either instructors or students for selected species. An example of a simple key for a few wood macerations is presented in **Box 12**.

Some plant species develop **tyloses** (parenchyma cells that invade the lumens of tracheary elements) either naturally during ageing or as a response to pathogen invasion. Old stems of morning glory (*Ipomoea purpurea*) and Dutchman's pipe (*Aristolochia durior*) often show this feature.

Prepare thin transverse and longitundinal sections of either stem, and stain with Toluidine Blue O. Scan the sections for the largest diameter tracheary elements, some of which should have tyloses (Figs. 100, 101). Earlier stages in tylose formation show the multicellular nature of these structures (Fig. 102).

An experiment demonstrating that translocation occurs through dead cells in the xylem is outlined in **Box 13**.

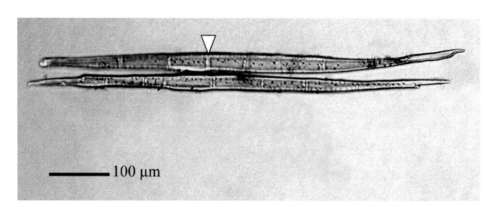

Figure 98. Septate fibres from macerated xylem of grape (*Vitis* sp.). Cross walls (septa) are evident (arrowhead). Viewed with partial polarizing microscopy.

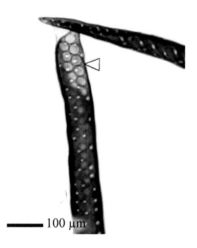

Figure 99. Vessel member from macerated xylem of ephedra (*Ephedra* sp.) showing the foraminate (ephedroid) perforation plate (arrowhead). Stained with TBO.

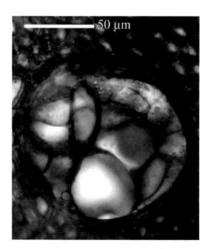

Figure 100. Transverse section of Dutchman's pipe (*Aristolochia durior*) stem stained with TBO showing tyloses within a large vessel element.

Figure 101. Longitudinal section of s (*Aristolochia durior*) stem stained with TBO showing tyloses within large vessel elements. Note that the tylose in the vessel member in the lower left corner has developed a secondary cell wall with pits.

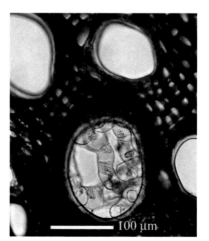

Figure 102. Transverse section of morning glory (*Ipomoea purpurea*) stem with developing tyloses within a large vessel element. Stained with TBO. Photo courtesy of Scott Liddycoat.

Box 12. Sample key to macerated xylem

I. Vessel members absent; tracheids with large circular bordered pits............... *Pinus* (pine)

II. Vessel members present; tracheids with less obvious bordered pits

    1a. Perforation plates simple

        2a. Most vessel members short and very wide (barrel shaped) with mainly transverse end walls ....*Gymnocladus* (Kentucky coffee tree)

        2b. Vessel members longer (cylindrical) with oblique end walls

            3a. Fibres septate... *Vitis* (grape)

            3b. Fibres non-septate

                4a. Vessel members with distinct groups of pits in lateral walls..... *Salix* (willow)

                4b. Vessel members with evenly distributed pits in lateral walls

                    5a. Vessel members often with an additional helical secondary cell wall laid down on the pitted secondary cell wall .... *Tilia* (basswood)

                    5b. Vessel members without an additional secondary cell wall ....*Asimina* (pawpaw)

    1b. Perforation plates multiperforate

        6a. Perforation plates scalariform...................................*Betula* (birch)

        6b. Perforation plates with a series of circular openings ....*Ephedra* (ephedra)

Box 13. Translocation through the xylem

One of the simplest and most conclusive ways of ascertaining whether a substance is being transported by the xylem or the phloem of plants is to determine whether the substance can be transported through a region of dead tissue. In this experiment, trisodium 3-hydroxy-5,8,10-pyrenetrisulfonate (PTS) is used to demonstrate transport through the xylem in a killed zone of *Phaseolus vulgaris* (bean) stems.

Select three healthy bean plants that have formed several sets of leaves. Prepare two test tubes or small beakers containing about 10 mL of 0.1% PTS dye and two tubes containing water. Excise three plants by cutting the stems below the lowest leaves. Place one in a tube with PTS and the other two in tubes with water. Using one of the shoots in water, follow the method of killing a localized zone on the stem (steam-ringing) as illustrated in **Figure 103**. After making the steam ring, transfer the shoot to the second tube with PTS. The shoot remaining in water is the control not treated with PTS. Place the tubes with the excised plants under lights to stimulate transpiration for at least one hour. Determine the distribution of the dye within the plants by means of a long-wave ultraviolet lamp. Compare the appearance of the plants placed in PTS with the one that was placed in water **(see Figure 104)**.

A similar experiment can be done using a solution of safranin O or other water-soluble dyes instead of PTS if a long-wave UV light is not available. However, the results are less dramatic, since it is more difficult to locate these dyes in the plant.

<u>NOTE</u>: Protective glasses must be worn when viewing the plants with UV light.

Figures 103 (a,b,c,d). Bean (*Phaseolus vulgaris*) plant showing method used to kill a localized region of the stem (known as steam-ringing) with hot water. The latter is easier to control than steam. Photos by Ryan Geil and Cameron Wagg.

Figure 103a. The stem is cut at the base and a piece of plastic tubing (arrow) is quickly slid up the stem before the cut end is placed into water. A small amount of absorbent cotton (arrowhead) is tied around the stem and twisted so that it is pointed. A twist tie (double arrowhead) is added to keep the plastic tubing in place later.

Figure 103b. Hot water (near boiling) is dripped onto the cotton for a few minutes to kill the cells in the region beneath the cotton. The plant can be held at an angle so that the hot water drips away from the stem. It is important to support the stem during and after this procedure to prevent bending and damage to the xylem.

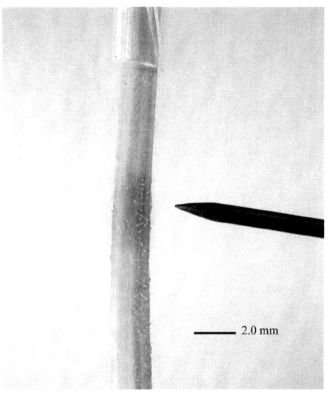

Figure 103c. The pointer indicates the darker green area in which cells have been killed. The cotton was removed revealing the region killed by hot water, but during the experiment the cotton should remain in place or the area should be coated with lanolin or stop-cock grease to prevent drying.

Figure 103d.
The plastic tubing is slipped over the wounded region to prevent bending of the stem.

Figure 104. Bean (*Phaseolus vulgaris*). Top row shows leaves from shoots that had been excised and placed in water, excised and placed in PTS dye, and excised and steam ringed before placing in PTS. Bottom row shows the underside of corresponding shoots viewed with UV light. PTS dye was transported equally well in shoots that had been steam-ringed or placed in the dye without steam-ringing. Because the tracheary elements of the xylem that transport the PTS are dead cells, killing a region of the stem does not affect its movement. Photos by Ryan Geil.

Phloem

Phloem is a complex tissue mainly involved in conducting photosynthates (sugars produced through photosynthesis) within the plant **(Box 14)**. Unlike xylem, it is a difficult tissue to study from fresh material. Despite this problem, a number of suggestions are presented to illustrate components of phloem. Very thin sections are required to visualize the small, thin-walled cells of this tissue.

Stems of sunflower provide very good material to study the basic organization of phloem tissue. Prepare thin transverse sections several internodes below the shoot apex and stain them with Toluidine Blue O. Locate the vascular bundles, each with phloem towards the cortex. Locate **sieve tube members**, parenchyma cells (some of which are **companion cells**), and the bundle cap of fibres (Fig. 105).

Stems of corn (*Zea mays*) have vascular bundles with very pronounced sieve tube members and companion cells that can be demonstrated in transverse sections stained with Toluidine Blue O (Fig. 106).

To study the phloem in more detail, stems of cucumber (*Cucumis sativus*) or other members of the Cucurbitaceae (e.g., squash) can be used. These plants have very wide-diameter sieve tube members so that the **simple sieve plate** can be demonstrated.

Prepare thin transverse sections of a stem, stain in Toluidine Blue O, and locate the vascular bundles using the 10× objective. The stem vascular bundles of members of the Cucurbitaceae have phloem located on both sides of the primary xylem **(bicollateral bundles)** evident in Fig. 107. Using the 40× objective, scan the phloem locating sieve tube members, companion cells, and other parenchyma cells (Fig. 108).

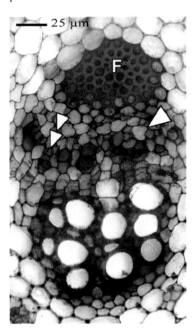
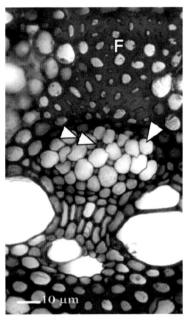
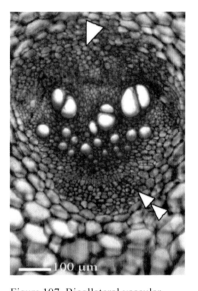

Figure 106. Vascular bundle from transverse section of corn (*Zea mays*) stem stained with TBO showing sieve tube members (arrowhead), companion cells (double arrowhead), and fibres (F) in the phloem.

Figure 105. Vascular bundle from a transverse section of sunflower (*Helianthus annuus*) stem stained with TBO. Primary phloem consists of sieve tube members (arrowhead), parenchyma, companion cells (double arrowhead), and fibres (F).

Figure 107. Bicollateral vascular bundle from transervse section of cucumber (*Cucumis sativus*) stem stained with TBO showing external (arrowhead) and internal (double arrowhead) phloem.

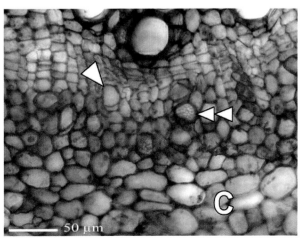

Figure 108. Transverse section of cucumber (*Cucumis sativus*) stem stained with TBO showing sieve tube members (arrowhead) and parenchyma cells in the external phloem. Simple sieve plates (double arrowhead) are visible in some sieve tube members. A portion of the cortex (C) is evident.

Box 14. Phloem

All vascular plants develop **primary phloem** that differentiates from part of the **procambium (provascular tissue)**, and only plants undergoing secondary growth develop **secondary phloem** from the vascular cambium. Both types of phloem are complex, consisting of a number of cell types. The **sieve elements (SE)** are the photosynthate-conducting cells and, depending on the plant group, these may be either **sieve cells** or **sieve tube members**. Sieve cells have unspecialized sieve areas and lack sieve plates; these are found in the phloem of gymnopserms and seedless vascular plants.

A **sieve tube** (found in flowering plants) is formed when two or more sieve tube members align end to end. Sieve tube members are separated by **sieve plates** that may be simple **(SS)** or compound. These plates have open pores allowing movement of photosynthates along the sieve tube. In addition to sieve elements, various types of parenchyma cells and fibres may be present. Specialized parenchyma cells, **companion cells (CC)**, are associated with sieve tube members in flowering plants. **Fibres (F)** are often formed as support cells for this tissue.

The diagram on the right shows the result of cutting into the tissue. Since the sieve tubes are under pressure, making an incision will cause the contents to rush towards the cut (direction of flow indicated by arrows) and the protoplasts (P) will collapse. Solid materials within the collapsed protoplasts accumulate on the sieve plates, forming so-called **'slime plugs'(SP)** that are mainly proteinaceous.

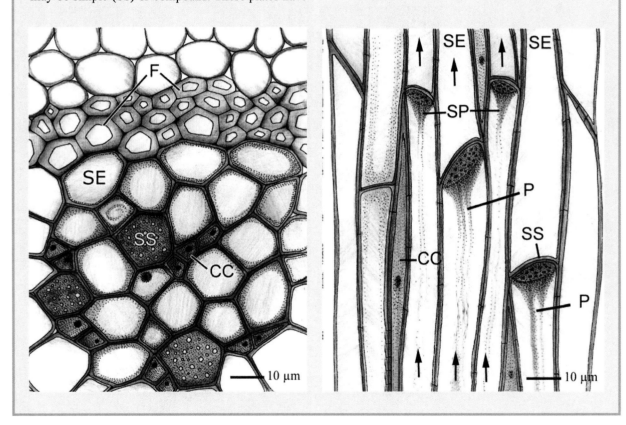

Several sections may have to be examined before simple sieve plates (Fig. 109) are located. In a section, it is common to find only parts of sieve plates.

The sieve plate pores form deposits of **callose** around them during wounding. The presence of callose can be demonstrated by staining tissue with aniline blue and viewing the stained samples with UV light using an epifluorescence microscope (Fig. 110).

Prepare thin longitudinal sections of a cucumber stem, stain with Toluidine Blue O, and locate sieve tube members and associated companion cells. Note that the sieve tube members have so-called **'slime plugs'** at their end because of the release of pressure within these cells during sectioning. This results in the movement of cell contents, including the **P-protein** toward the sieve plate (Fig. 111).

The proteinaceous nature of the slime plugs can be demonstrated by staining longitudinal sections with acid fuchsin (Fig. 112).

Longitudinal sections stained with aniline blue **(see Appendix 2)** and viewed with UV light by epifluorescence microscopy show a series of sieve tubes with callose at their ends where the sieve plates are located. These sieve elements also show small deposits of callose in the **lateral sieve areas** as well (Fig. 113).

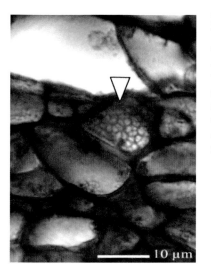

Figure 109. Transverse section of cucumber (*Cucumis sativus*) stem stained with TBO showing a simple sieve plate (arrowhead) with obvious pores in the end wall of a sieve tube member.

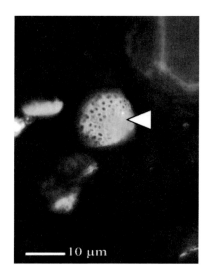

Figure 110. Transverse section of cucumber (*Cucumis sativus*) stem stained with aniline blue and viewed with UV light. A simple sieve plate with callose surrounding the pores is evident (arrowhead).

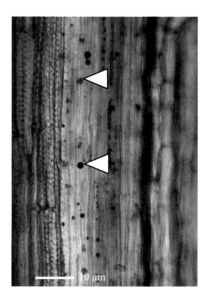

Figure 111. Longitudinal section of cucumber (*Cucumis sativus*) stem stained with TBO showing P-protein (arrowheads) at the ends of sieve tube members.

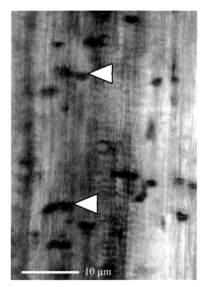

Figure 112. Longitudinal section of cucumber (*Cucumis sativus*) stem stained with acid fuchsin confirming the protein nature of the 'slime plugs' (arrowheads) at the ends of sieve tube members.

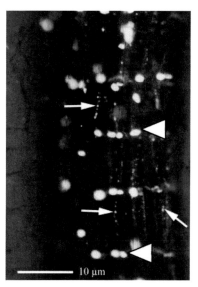

Figure 113. Longitudinal section of cucumber (*Cucumis sativus*) stem stained with aniline blue and viewed with UV light. Callose (arrowheads) surrounding pores in the sieve plates as well as in lateral sieve areas (arrows) is evident.

Chapter 6
Secretory structures

Plants synthesize a variety of compounds that may be excreted to the environment or stored within plant tissues. Specialized structures may be associated with the synthesis of these compounds. Secretory structures are categorized broadly into external and internal types.

External secretory structures

Glandular (secretory) trichomes

The most common external secretory structures are **glandular trichomes (hairs)**, epidermal outgrowths on aerial parts of plants. These are visible to the naked eye. Leaves, stems, and petals of plant species that are aromatic can provide good illustrations of the structural variation amongst these trichomes. A variety of compounds, the most common being monoterpenes, sesquiterpenes, and diterpenes, are secreted by glandular trichomes. Many of these are compounds that may protect the plant from insects and other herbivores.

Prepare thin sections of petioles, stems, or leaf blades of plant species such as geranium (*Pelargonium hortorum*), coleus (*Coleus blumei*), or other plants that are available, being careful not to handle the area to be sectioned. Mount the sections in water under a cover glass and observe them at various magnifications.

Some common plant species that are useful to illustrate secretory trichomes are listed, with illustrations of these structures.

- Petals of chrysanthemum
  (*Chrysanthemum morifolium*; Fig. 114)
- Petals of African violet
  (*Saintpaulia ionantha*; Fig. 115)
- Petioles and leaves of geranium
  (*Pelargonium hortorum*; Fig. 116)
- Petioles and leaves of mints (*Mentha* spp.) or sage
  (*Salvia officinalis*; Figs. 117, 118)

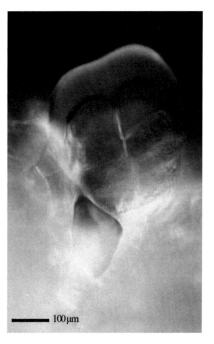

Figure 114. Secretory trichome from a petal of chrysanthemum (*Chrysanthemum morifolium*).

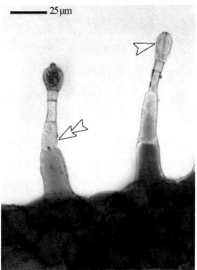

Figure 115. Secretory trichome from petal of African violet (*Saintpaulia ionantha*) showing head (arrowhead) and stalk cells (double arrowhead).

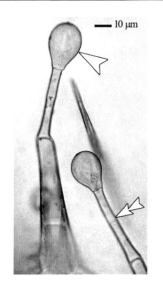

Figure 116. Secretory trichomes from geranium (*Pelargonium hortorum*) leaf showing head (arrowhead) and stalk cells (double arrowhead).

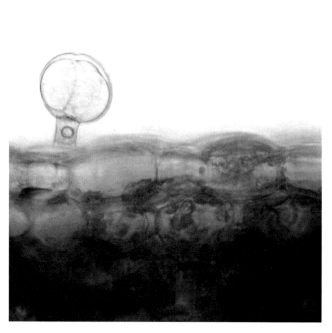

Figure 117. Secretory trichome with a short stalk from a sage (*Salvia officinalis*) leaf.

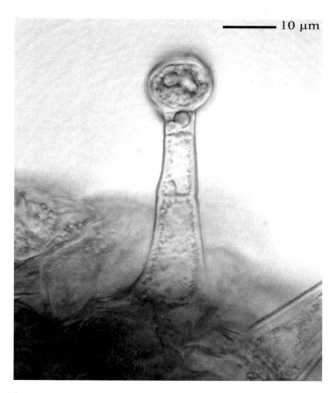

Figure 118. Secretory trichome with a long stalk from a sage (*Salvia officinalis*) leaf.

Tissue printing of whole leaves to demonstrate secretory trichomes

A simple method can be used to demonstrate the location and number of secretory trichomes on leaves **(refer to Box 4).**

- Obtain any member of the mint family (mint, thyme, sage, catnip) and remove young leaves, being careful not to damage the trichomes on the surface.

- Place a piece of nitrocellulose membrane on top of a few layers of Whatman filter paper in a Petri dish.

- Soak the nitrocellulose membrane in freshly prepared Tollen's reagent made by adding drops of 10% aqueous potassium hydroxide to 10% aqueous silver nitrate until a dark precipitate forms and then dissolving the silver oxide formed by adding drops of concentrated ammonium hydroxide.

This reagent should be prepared in a fume hood; avoid contact with skin.

- Blot membrane with filter paper to remove excess reagent.

- Place a leaf onto the membrane with the upper surface down and place a glass microscope slide on the leaf. Press on the slide for 10–15 seconds. Remove the leaf and observe the membrane with a stereobinocular microscope. Dark spots on the membrane show the sites of secretory trichomes (Fig. 119).

Figure 119. Tissue print of a leaf of sage (*Salvia officinalis*). The dark spots are caused by the reaction of compounds released from secretory trichomes with Tollen's reagent, producing metallic silver.

Nectaries

Nectaries can be classified as floral (located on floral parts) and extrafloral (located on other plant parts). There is a tremendous variation in floral nectaries ranging from modified trichomes to rather large multicellular structures.

Canola (*Brassica napus*) flowers can be used to demonstrate an example of the latter. Select flowers (Fig. 120) and, using a stereobinocular microscope, locate the nectaries at the base of the stamens (Fig. 121). Nectaries can be removed from the flowers easily with fine forceps and mounted whole in water under a cover glass (Fig. 122). Nectar is secreted through numerous large, permanently open stomata that are prominent on the surface of the lobed nectaries (Fig. 123).

Figure 120. Flowers of canola (*Brassica napus*).

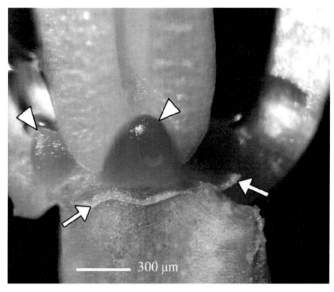

Figure 121. Nectaries (arrowheads) at the base of the stamens in a flower of canola (*Brassica napus*) from which the petals have been removed (arrows).

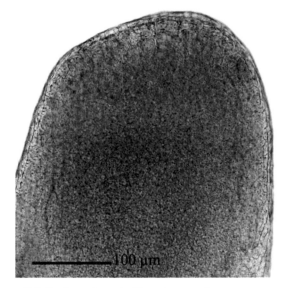

Figure 122. Surface of canola (*Brassica napus*) nectary.

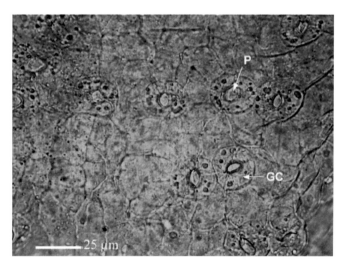

Figure 123. Higher magnification of the nectary surface of canola (*Brassica napus*) showing numerous stomata consisting of guard cells (GC) and pores (P) that remain open.

Petioles of passion flower (*Passiflora caerulea*) have large cup-shaped **extrafloral nectaries** (Fig. 124). These can be sectioned and stained with TBO to demonstrate the nectariferous region indicated by cells compactly arranged and with large nuclei (Fig. 125).

Hydathodes

These structures occur on the leaves of many species and are involved in a process called **guttation**, the discharge of water and dissolved ions. Hydathodes usually occur along the margins or at the tips of leaves and are characterized by abundant xylem leading to the hydathode, modified mesophyll cells, and several stomata that remain open. A large hydathode, illustrating these features, can be seen in Fig. 126, a cleared leaf of the aluminum plant (*Pilea cadierei*).

Internal secretory structures

Many plant species develop specialized cells or groups of cells within stems, leaves, roots, and fruits that synthesize a variety of compounds including resins, oils, latex, and tannins. A few examples follow.

Secretory ducts and cavities

Prepare thin transverse sections of pine (*Pinus* spp.) needles (leaves) and stain with either Toluidine Blue O or a Sudan dye. **Resin ducts** occur in various numbers depending on the pine species used. Large resin ducts in Scots pine (*Pinus sylvestris*) are located in the mesophyll (Fig. 127) and the resin they contain stains with Sudan dyes (Figs. 127, 128). Fibres surround the secretory (**epithelial**) cells (Fig. 128). In the wedge-shaped white pine (*Pinus strobus*) needles there are fewer resin ducts (Fig. 129) and the cells surrounding the epithelial cells are not as thick-walled (Fig. 130).

Prepare thin transverse sections of a sunflower (*Helianthus annuus*) stem and stain with TBO. Secretory ducts are present in the cortex (Fig. 131) and at higher magnification, epithelial cells are evident (Fig. 132).

The base (attachment to stalk) of banana (*Musa* sp.) fruits contains secretory ducts that are evident in unstained sections (Fig. 133).

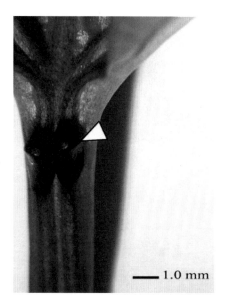

Figure 124. Extrafloral nectaries (arrowhead) on a petiole of a passion flower (*Passiflora caerulea*) leaf.

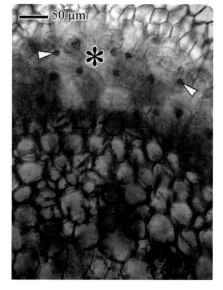

Figure 125. Section of extrafloral nectary of passion flower (*Passiflora caerulea*) stained with TBO showing the secretory surface (*). Cells in this region have large nuclei (arrowheads) and dense cytoplasm.

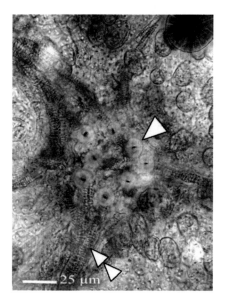

Figure 126. Top view of a hydathode from a cleared leaf of aluminum plant (*Pilea cadierei*) showing numerous stomata (arrowhead) on the surface and enlarged tracheary elements (double arrowhead) leading into the hydathode. Photo courtesy of Brita Ng.

Secretory Structures

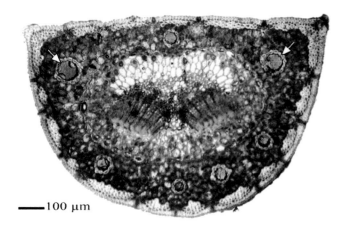

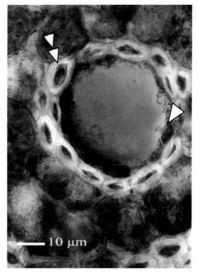

Figure 127. Transverse section of Scots pine (*Pinus sylvestris*) needle stained with Sudan dye. Resin ducts (arrows) are present in the mesophyll.

Figure 128. Resin duct in a transverse section of Scots pine (*Pinus sylvestris*) needle stained with a Sudan dye. The duct is filled with resin and is surrounded by epithelial cells (arrowhead) and sclerified parenchyma or fibres (double arrowhead).

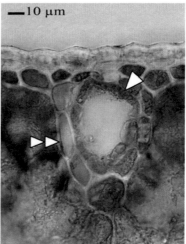

Figure 129. Transverse section of white pine (*Pinus strobus*) needle stained with TBO. A few resin ducts (arrows) are present.

Figure 130. Unstained transverse section of white pine (*Pinus strobus*) needle showing a resin duct surrounded by epithelial cells (arrowhead) and sclerified parenchyma or fibres (double arrowhead).

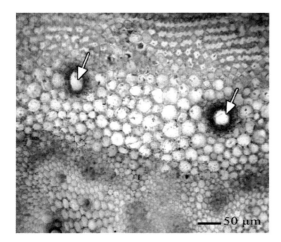

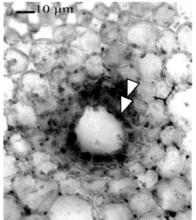

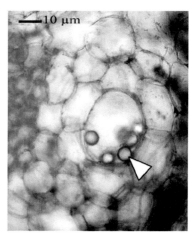

Figure 131. Secretory ducts (arrows) in a transverse section of sunflower (*Helianthus annuus*) stem stained with TBO.

Figure 132. Secretory duct in a transverse section of sunflower (*Helianthus annuus*) stem stained with TBO. The duct is surrounded by epithelial cells (double arrowhead) that contain chloroplasts.

Figure 133. Unstained section of the base of a banana (*Musa* sp.) fruit. Oil droplets (arrowhead) are present in the secretory duct.

Prepare thin paradermal or fairly thick transverse sections of orange (*Citrus sinensis*) peel and mount in water. Several **secretory cavities** (i.e., spaces formed by the breakdown of cells) will be evident (Fig. 134). Stain some sections with Sudan. Contents within the lumen and surrounding cells of the cavities that stain red with Sudan dyes (Fig. 135) will be visible. Although there is some controversy as to whether or not these cavities have epithelial cells, some sections do show what appear to be secretory cells surrounding the lumen (Fig. 136).

Celery (*Apium graveolens*) petioles have narrow **secretory ducts** containing compounds that give this vegetable its characteristic odor and taste. Transverse sections stained with Toluidine Blue O show the location of the secretory ducts consisting of epithelial cells and a lumen (Fig. 137). The location and number of these within the celery petiole can be demonstrated with a tissue printing **(Box 4)** method.

- Place several layers of Whatman filter paper in a Petri dish and place a piece of nitrocellulose membrane over these.
- Cut a section (about 5 cm long) from a petiole and press the cut end onto the membrane. Do this a number of times on various regions of the membrane.
- Let the membrane dry briefly and then observe it with a hand-held UV light source in the dark. The sites of secretory ducts will fluoresce because of the presence of the secretory compounds transferred from the ducts to the membrane (Fig. 138).

Warning: Protective goggles must be worn when viewing the membrane under UV light. Skin should not have prolonged exposure to UV light.

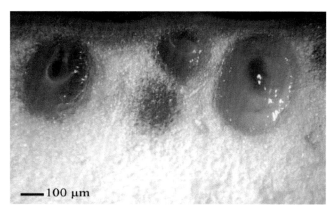

Figure 134. Thick transverse section of orange (*Citrus sinensis*) peel showing several secretory cavities, the source of oils.

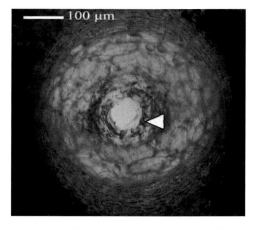

Figure 135. Paradermal section of orange (*Citrus sinensis*) peel stained with Sudan dye showing droplets of oil (arrowhead) in cells surrounding the cavity.

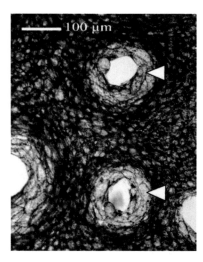

Figure 136. Paradermal section of orange (*Citrus sinensis*) peel showing secretory cavities with what appear to be intact epithelial cells (arrowheads).

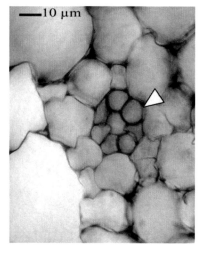

Figure 137. Transverse section of celery (*Apium graveolens*) petiole stained with TBO showing a secretory duct lined with epithelial cells (arrowhead).

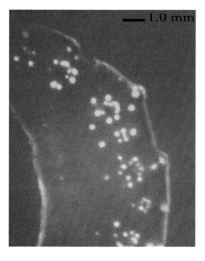

Figure 138. Tissue blot of celery (*Apium graveolens*) petiole viewed under UV light showing the distribution of secretory ducts.

Secretory Structures

Laticifers

Many plants produce **latex** that is synthesized and stored in elongated individual cells or files of cells called **laticifers**. These structures can have various forms and their contents (latex) can be of a variety of colours. Frequently latex is whitish and 'leaks out' of organs when they are cut.

Stems of crown-of-thorns (*Euphorbia splendens*) have many laticifers that can be viewed in both transverse and longitudinal sections. Remove the spines from a stem segment and prepare thin transverse and longitudinal sections. Stain some with Toluidine Blue O and others with I_2KI for starch.

Laticifers in transverse sections stained with Toludine Blue O will appear as thick-walled (pink–purple coloured) cells in the cortex and pith (Figs. 139, 140).

In longitudinal section, laticifers appear as elongate thick-walled structures that ramify through the tissues (Fig. 141). The starch grains within these laticifers have unique 'dog bone' (dumbbell) shapes; starch grains in adjacent parenchyma cells have the more common oval shapes (Fig. 142).

Warning: Do not expose skin to latex. Some individuals may have an allergic reaction.

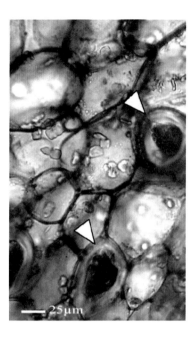

Figure 139. Transverse section of crown-of-thorns (*Euphorbia splendens*) stem stained with TBO showing thick-walled laticifers (arrowheads) in the cortex. The dark contents are latex.

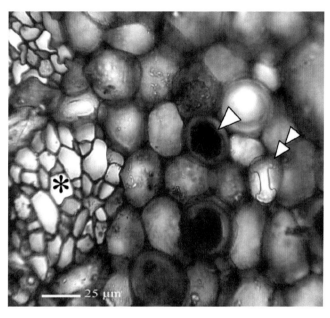

Figure 140. Transverse section of crown-of-thorns (*Euphorbia splendens*) stem stained with TBO showing thick-walled laticifers (arrowhead) in the cortex adjacent to the phloem (∗) A 'dog-bone shaped' starch grain (double arrowhead) within a laticifer is evident.

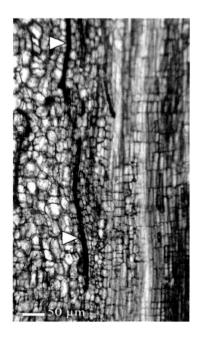

Figure 141. Longitudinal section of crown-of-thorns (*Euphorbia splendens*) stem stained with TBO showing distribution of laticifers (arrowheads).

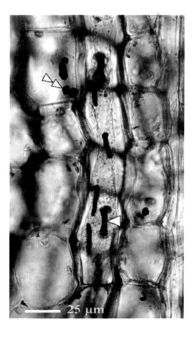

Figure 142. Longitudinal section of crown-of-thorns (*Euphorbia splendens*) stem stained with I_2KI showing the characteristic 'dog-bone shaped' starch grains (arrowhead) in a laticifer. Starch grains in adjacent cells (double arrowhead) are spherical.

Chapter 7
Roots

Most roots grow underground and thus are not as easy to study as aerial parts of plants. However, there are many ways that living material can be used to supplement the usual permanent slides used in courses.

Root caps

Germinated caryopses (kernels) of corn (*Zea mays*) can be used to show root cap structure and production of mucilage. Caryopses should be germinated on moist filter paper in a Petri dish and grown until the **radicle (young primary root)** is a few centimetres in length. Pick up a germinated seedling, cut the radicle from the caryopsis, and place in a drop of water on a microscope slide. Add a drop of TBO and observe under low magnification with a microscope. The secreted **mucilage** with sloughed peripheral root cap cells (**border cells**) will be evident (Figs. 143, 144).

Root hairs

A method for producing material for studying living root hairs is outlined in **Box 15.** Obtain a grass seedling grown by the method in **Box 15**, remove the strip of paper towel, and blot any water from the surfaces of the cover glass and slide with filter paper or any absorbent tissue. Observe the root with a compound microscope. The root apex is covered by a small root cap and some of the peripheral cells will be in the process of being sloughed (Fig. 145). Beginning at the root apex, follow the development of root hairs as you move back from the apex. Small **root hair papillae** (Fig. 146) are initiated from epidermal cells close to the root apex and these elongate by **tip growth**. Elongated hairs can be seen in older parts of the root, i.e., further back from the root apex (Fig. 147). Choose an elongated root hair (Fig. 148) and observe the streaming in the parietal cytoplasm along the cell wall and in transvacuolar strands.

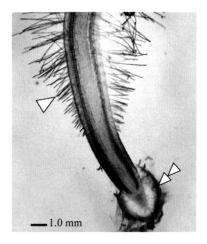

Figure 143. Root tip of corn (*Zea mays*) seedling mounted fresh and stained with TBO. Developing root hairs (arrowhead) and mucilage surrounding the root cap (double arrowhead) are evident.

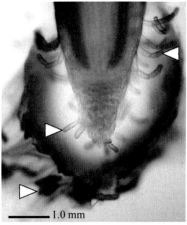

Figure 144. Higher magnification of corn (*Zea mays*) root tip stained with TBO. Sloughed root cap cells (arrowheads) are evident in the mucilage secreted around the root tip.

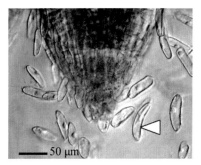

Figure 145. Apex of *Festuca* sp. root showing sloughed root cap cells (arrowhead). Viewed with Nomarski interference contrast microscopy.

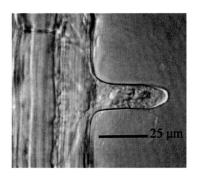

Figure 146. Early stage in development of a root hair in a seedling of *Festuca* sp. viewed with Nomarski interference contrast microscopy.

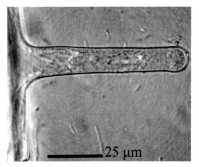

Figure 147. Elongated root hair in a seedling of *Festuca* sp. viewed with Nomarski interference contrast microscopy.

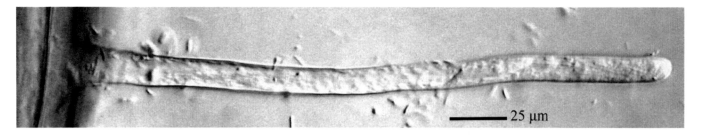

Figure 148. Mature root hair in a seedling of *Festuca* sp. viewed with Nomarski interference contrast microscopy.

> ### Box 15. Growing grass seedlings for observation of root hairs and lateral roots
>
> Seeds (caryopses) of most grass species can be used for this method. Lawn seed that can be purchased at local hardware stores or nurseries is excellent material for this exercise.
>
> 1. Germinate seeds on moist filter paper and let seedlings grow until the radicle (primary root) is 1–2 cm in length and a shoot has formed.
> 2. Transfer seedlings to water on a microscope slide and cover the radicle with a long cover glass (22 mm × 50 mm works well), leaving the shoot exposed.
> 3. Place a strip of paper toweling over the cover glass.
> 4. Stand the slides in a coplin jar or small beaker containing a small amount of water and cover so that the seedling does not dry out. Allow the seedlings to grow for several days.
>
>
>
> Photo courtesy of Ryan Geil

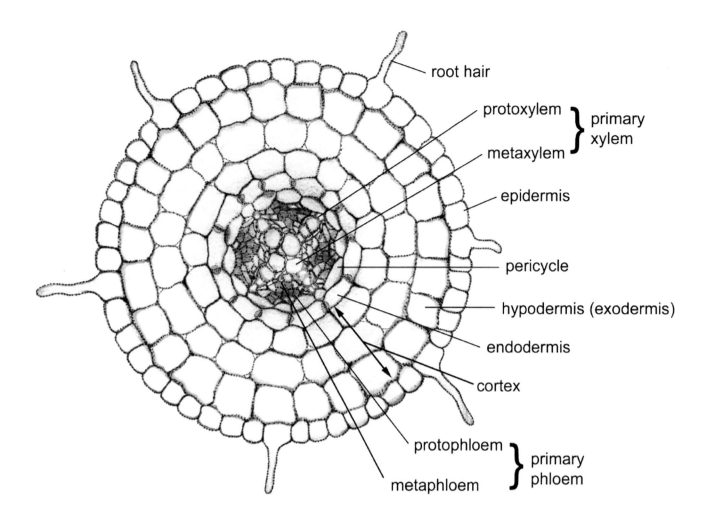

Figure 149. Diagram showing organization of primary tissues in a dicotyledonous species root.

Primary tissues in roots

The organization of primary tissues in roots varies, particularly in the amount of **primary phloem** and **primary xylem** formed, and whether or not the center of the root differentiates as **pith**. Figure 149 illustrates the primary tissues present in the majority of roots; it serves as a guide for the following examples. If a series of sections is needed for various stains, the method outlined in **Chapter 2** can be used.

Buttercup (*Ranunculus* spp.) roots are useful to demonstrate primary tissue organization in a dicotyledenous plant species since, in most species of this genus, very little secondary growth occurs. Obtain plants from the field, keeping the roots moist. Prepare transverse sections and stain with Toluidine Blue O. The organization of the primary tissues is similar to that illustrated in the diagram (Fig. 149). *Ranunculus flabellaris*, the species illustrated in Fig. 150, grows in wet habitats, resulting in the formation of large air spaces (**aerenchyma**) for the conduction of oxygen to the root from the shoot. A higher magnification of the vascular cylinder shows the **tetrarch** arrangement (i.e., four protoxylem poles) of primary xylem with alternating regions of primary phloem, the **pericycle**, **endodermis**, and parenchyma cells in the **cortex** (Fig. 151).

Sections stained with phloroglucinol-HCl show that the **Casparian band** and cell walls in the endodermis (**Box 16**), as well as the tracheary elements of the xylem, are lignified (Fig. 152). Sections stained with Sudan dyes show the positive staining for suberin in the Casparian band and in the walls **suberin lamellae** of endodermal cell walls (Fig. 153).

Roots of the dicotyledonous species tomato (*Solanum lycopersicum*) show the variation in arrangement of primary xylem and phloem within roots of a single species. Diarch (Fig. 154), triarch (Fig. 155), and tetrarch (Fig. 156) arrangements are common. Figure 156 also shows that the protoxylem matures before the metaxylem in roots.

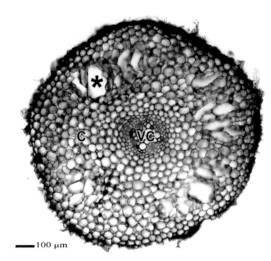

Figure 150. Transverse section of buttercup (*Ranunculus flabellaris*) root stained with TBO showing organization of primary tissues. C = cortex; VC = vascular cylinder. Large air spaces (∗) are present in the cortex, an indication that this buttercup species grows in very wet soil.

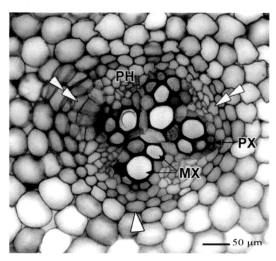

Figure 151. Transverse section of buttercup (*Ranunculus flabellaris*) root stained with TBO showing organization of the vascular cylinder. PX = protoxylem; MX = metaxylem; PH = phloem; arrowhead = endodermis; double arrowheads = pericycle.

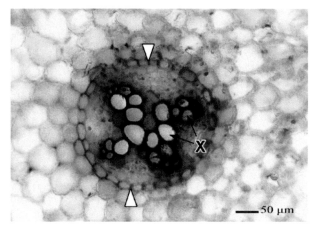

Figure 152. Transverse section of buttercup (*Ranunculus flabellaris*) root stained with phloroglucinol-HCl showing that the xylem tracheary elements (X) and the endodermal cell walls including the Casparian bands (arrowheads) contain lignin.

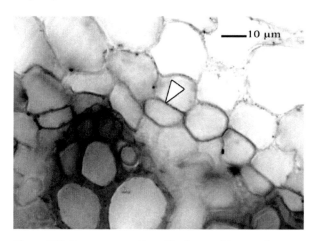

Figure 153. Transverse section of buttercup (*Ranunculus flabellaris*) root stained with a Sudan dye showing the presence of suberin in endodermal cell walls (arrowhead).

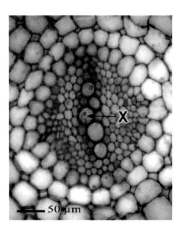

Figure 154. Transverse section of tomato (*Solanum lycopersicum*) root stained with TBO showing diarch primary xylem (X).

Figure 155. Transverse section of tomato (*Solanum lycopersicum*) root stained with TBO showing triarch primary xylem (X).

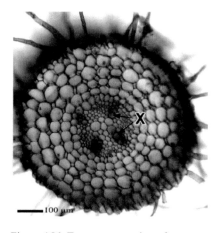

Figure 156. Transverse section of tomato (*Solanum lycopersicum*) root stained with TBO showing tetrarch primary xylem (X). Only protoxylem has stained, metaxylem has not matured.

Roots of corn, a monocotyledonous species, are easy to section and illustrate primary tissue organization very clearly. Germinate caryopses (kernals) in moist vermiculite and allow seedlings to grow until several roots have formed. Select roots that have the widest diameter (as these are the easiest to section), prepare transverse sections, and stain with Toluidine Blue O. A **uniseriate epidermis** (some of the cells of which have formed root hairs), a **cortex,** and **vascular cylinder** are evident at low magnification (Fig. 157). At higher magnification, the **exodermis** consisting of tightly packed cells, endodermis, primary xylem, and primary phloem are evident (Fig. 158). **Protoxylem**, large mature early **metaxylem vessels** and very large immature late metaxylem vessels, and **pith** are shown in Figures 158 and 159.

Exodermis and endodermis

Roots of all vascular plant species develop an endodermis, the inner layer of cortex, and most species of flowering plants develop an exodermis, one or more specialized layers of outer cortical cells. These cells have wall modifications as outlined in **Boxes 16 and 17**. Transverse sections of corn roots stained with berberine and viewed with blue or UV light with an epifluorescence microscope demonstrate the presence of **Casparian bands** in both the exodermis and endodermis (Fig. 160).

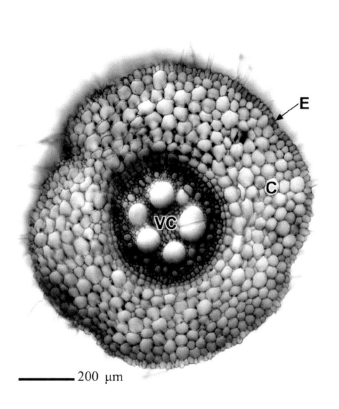

Figure 157. Transverse section of corn (*Zea mays*) root stained with TBO showing primary tissue organization. E = epidermis; C = cortex; VC = vascular cylinder.

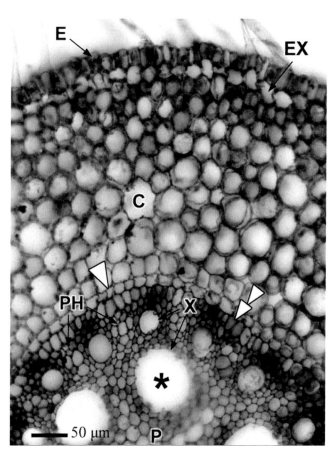

Figure 158. Transverse section of corn (*Zea mays*) root stained with TBO viewed at higher magnification. E = epidermis; EX = exodermis; C = cortex; arrowhead = endodermis; double arrowhead = pericycle; X = xylem; PH = phloem; P = pith. Large immature metaxylem vessels (∗) are evident.

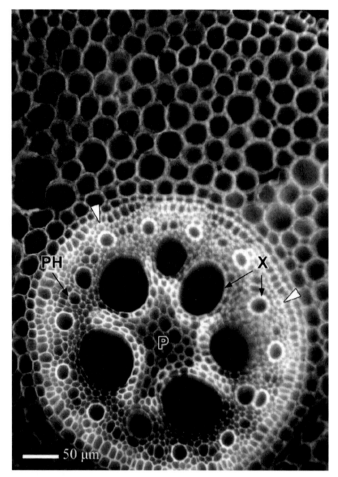
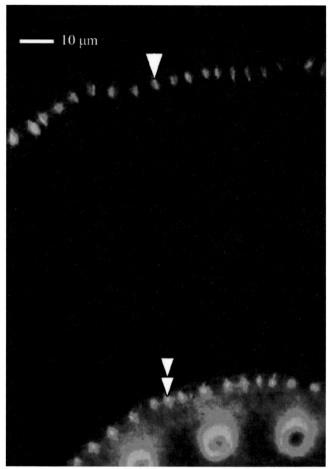

Figure 159. Transverse section of corn (*Zea mays*) root stained with Cellufluor and viewed with UV light showing the alternating arrangement of xylem (X) and phloem (PH) poles. Roots such as these that have many protoxylem poles (arrowheads) are termed polyarch. The center of the root is occupied by pith (P).

Figure 160. Transverse section of corn (*Zea mays*) root stained with berberine and examined with blue light. Casparian bands are evident in the exodermis (arrowhead) and the endodermis (double arrowhead).

Adventitious roots formed from bulbs of the monocotyledonous onion (*Allium cepa*) are easy to obtain and hand-section. Remove the brown scale-like outer leaves from onion bulbs and place them in moist vermiculite or over water in a container just wide enough to support the bulb. Allow adventitious roots to form. Remove roots and prepare thin transverse sections. These can be stained in a number of ways. **Epidermis, exodermis (see Box 17)** consisting of two cell types (dimorphic), **cortex, endodermis,** and **vascular cylinder** (consisting of **pericycle**, primary xylem and primary phloem) constitute the primary tissues of this root (Fig. 161). The transverse section in Fig. 161 has been stained with Cellufluor, a fluorescent dye that binds to cellulose in the cell wall. Sections stained with berberine and viewed with UV light using an epifluorescence microscope show the Casparian bands in the endodermis (Fig. 162) and in the exodermis (Fig. 163).

Sections through older roots stained with Sudan red 7B (fat red) show the suberized walls of the epidermis, exodermis, and endodermis (Fig. 164). A higher magnification of an even older section of root shows the arrangement of the primary xylem and phloem clearly and a late stage in development of the endodermis (Fig. 165). Figure 166 shows the arrangement of short and long cells of the exodermis, while Figure 167, a section stained with neutral red, shows that at this level in the root, the short cells accumulate the stain, indicating that they are alive **(for details see Appendix 5).**

Roots

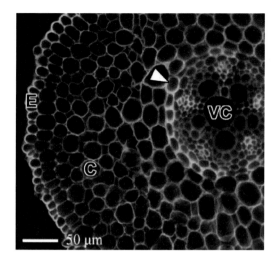

Figure 161. Transverse section of onion (*Allium cepa*) root stained with Cellufluor and viewed with UV light showing organization of primary tissues. E = epidermis; C = cortex; arrowhead = endodermis; VC = vascular cylinder.

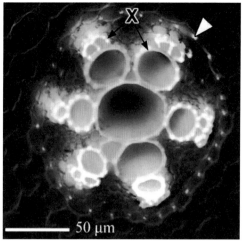

Figure 162. Transverse section of onion (*Allium cepa*) root stained with berberine and viewed with UV light showing organization of primary xylem (X) and the presence of Casparian bands (arrowhead) in the endodermis.

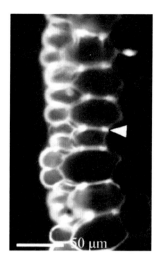

Figure 163. Transverse section of onion (*Allium cepa*) root stained with berberine and viewed with UV light showing Casparian bands (arrowhead) in the exodermis.

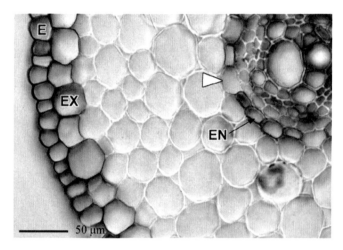

Figure 164. Transverse section of onion (*Allium cepa*) root stained with Sudan dye. Suberin is present in the cell walls of the epidermis (E), exodermis (EX), and endodermis (EN). A passage cell, i.e., without suberin lamellae (arrowhead), is evident in the endodermis.

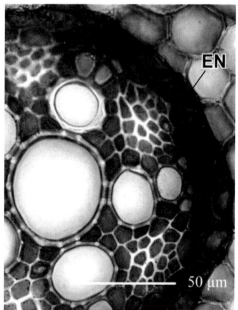

Figure 165. Transverse section of onion (*Allium cepa*) root stained with Sudan dye. Walls of endodermal cells (EN) are highly suberized and asymmetrically thickened.

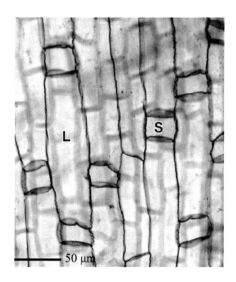

Figure 166. Paradermal section, i.e., parallel to the surface, of onion (*Allium cepa*) root stained with chlorazol black E showing the arrangement of short (S) and long (L) cells of the exodermis. An exodermis having these two cell types is referred to as dimorphic. Photo courtesy of Susan Benedetto.

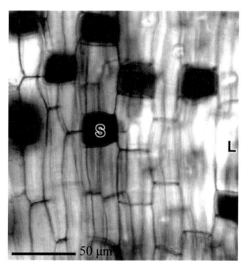

Figure 167. Paradermal section of onion (*Allium cepa*) root stained with neutral red showing that in the exodermis, the short cells (S) have permeable walls and retain their protoplasts longer than the long cells (L). Photo courtesy of Elida Kafarowski.

Box 16. Endodermis

Roots, along with mycorrhizal fungi associated with most species, are involved in taking up mineral ions from the soil solution and transporting these to the shoot system. The **endodermis (EN)** plays two important roles in this process. The endodermis is the innermost layer of cells in the cortex (C) and, thus, surrounds the vascular cylinder. Endodermal cells have **Casparian bands (CB)**, deposits of lignin and suberin that infiltrate a band-like region in their anticlinal, primary walls (see diagrams below). Casparian bands greatly impede the passage of ions from the soil solution to the vascular cylinder through the cell walls. This means that the ions that do enter the xylem are under the control of membrane proteins in the epidermis and cortex. Once inside the vascular cylinder, the ions to be transported to the shoot are moved from the cells' cytoplasm to their cell walls, and the Casparian band prevents these ions from leaking out of the vascular cylinder through the cell walls. The endodermis, especially in the roots of many monocotyledonous species, develops **suberin lamellae**, i.e., thin layers of suberin inside the cell's primary walls. An additional thick wall layer (typically asymmetric) containing cellulose, and often lignin and suberin, may then be laid down inside the suberin lamella.

Suberin and lignin display **primary fluorescence** when viewed with UV light. This fluorescence can be augmented by staining sections with the fluorochrome, berberine **(see Appendix 2)**.

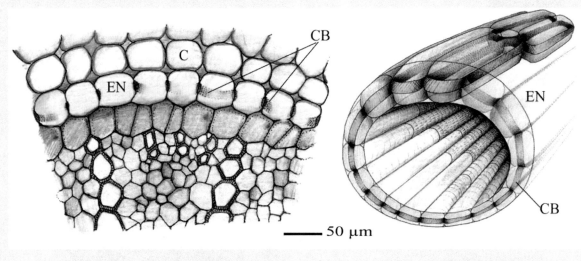

Diagram of root cross section   3-D view of endodermal cylinder

Box 17. Exodermis

In angiosperm plants, the cortical layer next to the epidermis (E) is usually modified to form an exodermis. Its cells are distinct from those of the central cortex (CC). This layer resembles the endodermis in that it possesses Casparian bands (yellow walls in diagrams of cross sections). Unlike the endodermis, in which the Casparian bands form within a few millimetres of the tip, the Casparian bands of the exodermis may form several centimetres from the root tip. Suberin lamellae (red in cross sections) encircle the cells with the exception of short cells (SC).

There are two main types of exodermis: uniform and dimorphic. As seen in face view (from a **paradermal section** of the root) the cells of the uniform exodermis are similar in length and all develop suberin lamellae (shaded pink in the diagram below). In contrast, the dimorphic exodermis is made up of long cells and short cells; suberin lamellae form first in the long cells. In nature, the root epidermis often dies leaving the exodermis, with its protective suberized walls, as the outermost layer of the root. The exodermis can be visualized by clearing and staining whole roots with chlorazol black E, or staining sections with Sudan dyes for suberin lamellae, or with berberine-aniline blue for Casparian bands.

UNIFORM

Transverse Sections

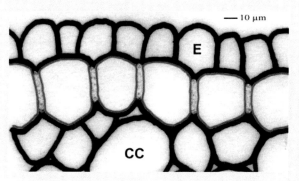

Paradermal Sections

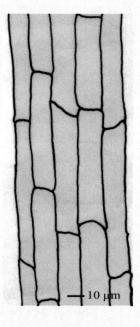

DIMORPHIC

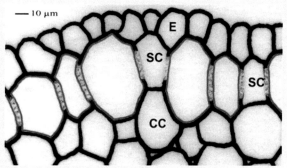

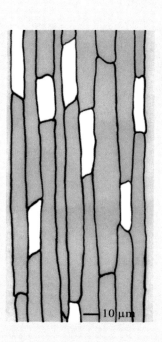

Roots with phi thickenings

Several plant species develop **phi thickenings** (so named because they resemble the Greek letter Φ when viewed in transverse section) in one or more layers of cortical cells. These are lignified thickenings that probably function for support of this region of the root.

Geranium (*Pelargonium hortorum*)

Cuttings of geranium should be rooted in moist vermiculite to produce adventitious roots. Prepare thin, transverse sections of these and stain with either Toluidine Blue O or phloroglucinol-HCl. Large phi thickenings occur in cortical cells adjacent to the epidermis, while smaller phi thickenings occur in other cortical cells (Fig. 168). The continuous nature of the thickenings can be demonstrated in longitudinal sections stained with phloroglucinol-HCl (Fig. 169). The phi thickenings show autofluorescence when sections are viewed with UV light (Fig. 170) and stain blue with Toluidine Blue O (Fig. 171), indicating the presence of lignin.

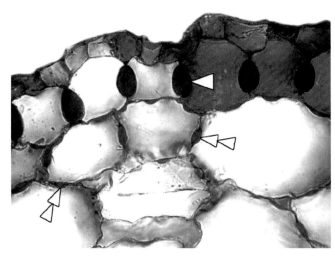

Figure 168. Transverse section of geranium (*Pelargonium hortorum*) root stained with phloroglucinol-HCl. Large, lignified phi thickenings (arrowhead) are present in the hypodermis and smaller phi thickenings (double arrowheads) are present in other cortical cells. Photo courtesy of Chris J. Meyer.

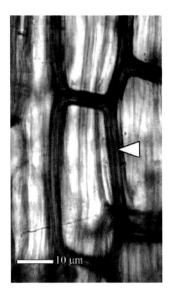

Figure 169. Longitudinal section of geranium (*Pelargonium hortorum*) root stained with phloroglucinol-HCl. Lignified phi thickenings (arrowhead) are evident. Photo courtesy of Chris J. Meyer.

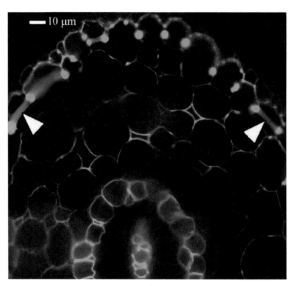

Figure 170. Unstained transverse section of geranium (*Pelargonium hortorum*) root viewed with UV light. The large phi thickenings, some appearing as bands on the cell surface in the hypodermis (arrowheads), are evident. Photo courtesy of Chris J. Meyer.

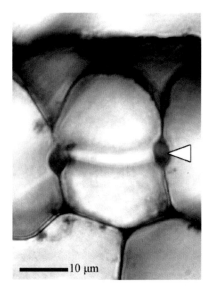

Figure 171. Transverse section of geranium (*Pelargonium hortorum*) root stained with TBO showing the lignified nature of phi thickenings (arrowhead). Photo courtesy of Becky Longland.

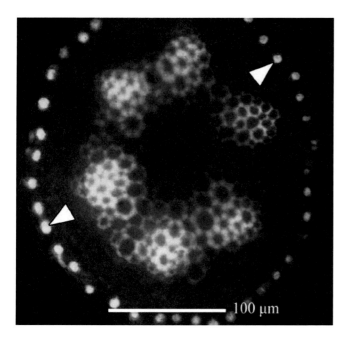

Figure 172. Unstained transverse section of apple (*Malus pumila*) root viewed with blue light showing large phi thickenings (arrowheads) in the cortical cells adjacent to the endodermis.

Figure 173. Unstained transverse section of eastern white-cedar (*Thuja occidentalis*) root viewed with UV light showing large phi thickenings (arrowhead) in cortical cells next to the endodermis and smaller phi thickenings (double arrowheads) throughout the cortex. Passage cells (PC) are evident in the endodermis.

Apple (*Malus pumila*)

Transverse sections of roots observed unstained with blue light show the pronounced phi thickenings in cortical cells adjacent to the endodermis (Fig. 172).

Eastern white-cedar (*Thuja occidentalis*)

Unstained transverse sections viewed with UV light show large phi thickenings in the inner cortex and numerous smaller thickenings throughout the cortex (Fig. 173). This figure also shows the **passage cells** (i.e., cells without suberin lamellae) in the endodermis.

Lateral roots

Grass seedlings grown as indicated in **Box 15** can be used to demonstrate the development of lateral roots. Use grass seedlings that have been allowed to grow for several days to a week and follow the initiation and outgrowth of lateral roots from the periphery of the vascular cylinder (Figs. 174–176).

A method for examining the development of lateral root primordia without sectioning roots is outlined in **Box 18**. Roots of most species can be used for this exercise but tree roots often are difficult to clear because of the presence of condensed tannins.

Seedlings of corn (*Zea mays*) grown in moist vermiculite are useful to show the development of lateral roots from sections. Prepare transverse sections of older roots beginning

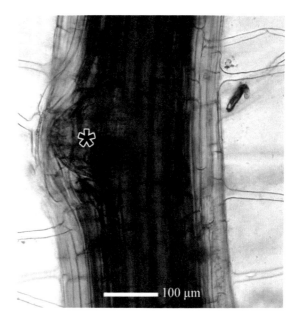

Figure 174. Lateral root primordium (∗) in a *Festuca* sp. seedling.

close to the apex and working back, stain with Toluidine Blue O, and observe the sites of lateral root initiation, and the disruption of exodermal and epidermal tissues during lateral root outgrowth (Fig. 177).

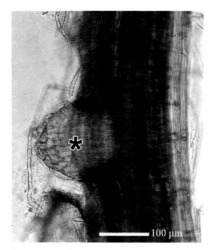

Figure 175. Lateral root primordium (*) that has just emerged from the parent root in a *Festuca* sp. seedling.

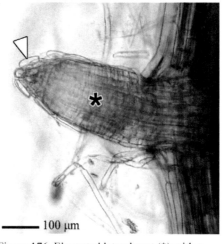

Figure 176. Elongated lateral root (*) with a root cap (arrowhead) in a *Festuca* sp. seedling.

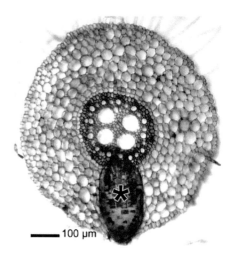

Figure 177. Transverse section of corn (*Zea mays*) root stained with TBO showing a lateral root primordium (∗).

Secondary growth in roots

Roots of a number of species including tomato (*Solanum lycopersicum*), soybean (*Glycine max*), and pea (*Pisum sativum*) from seedlings grown in moist vermiculite or hydroponics can be used to demonstrate the initiation of the **vascular cambium** and the subsequent formation of **secondary xylem** and **secondary phloem**. Prepare thin, transverse sections at various distances back from the root apex (best in the region where lateral roots have developed), stain in Toluidine Blue O, and determine the organization of primary tissues, the initiation of the vascular cambium (Fig. 178), and the formation of secondary xylem and secondary phloem (Figs. 179, 180). Figures 181a and 181b illustrate the changes that occur in a tetrarch root as the vascular cambium divides to form secondary xylem and secondary phloem. The pericycle divides to accommodate the increase in diameter of the root and also to form at least two layers. The endodermis and the cortical cells normally do not divide or divide for a short period of time and will be sloughed from the surface of the root. Cells derived from the pericycle will form a **phellogen (cork cambium)** that will give rise to derivatives forming the **phellem** and perhaps some **phelloderm**. Together, these layers are referred to as the **periderm** tissue. Details of this process are given in the section on secondary growth in stems.

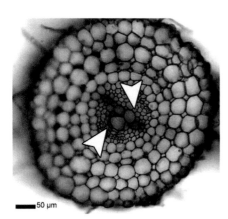

Figure 178. Transverse section of tomato (*Solanum lycopersicum*) root stained with TBO, showing sites of vascular cambium initiation (arrowheads).

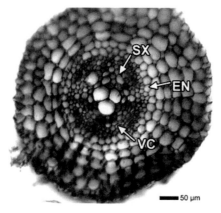

Figure 179. Transverse section of tomato (*Solanum lycopersicum*) root stained with TBO, showing secondary xylem (SX) formation. The endodermis (EN) and vascular cambium (VC) are evident.

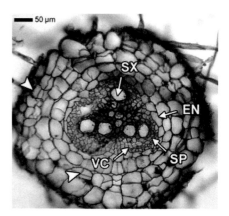

Figure 180. Transverse section of tomato (*Solanum lycopersicum*) root stained with TBO, showing secondary xylem (SX) and secondary phloem (SP) formation. The endodermis (EN), vascular cambium (VC), and divisions in the outer cortex (arrowheads) are evident.

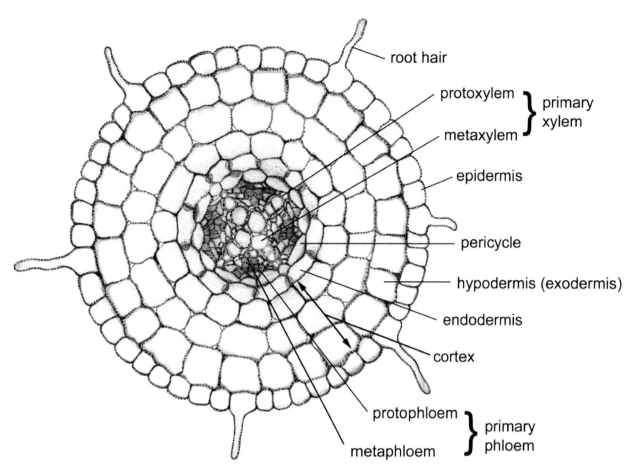

Figure 181a. Diagram showing organization of primary tissues in a dicotyledonous species root.

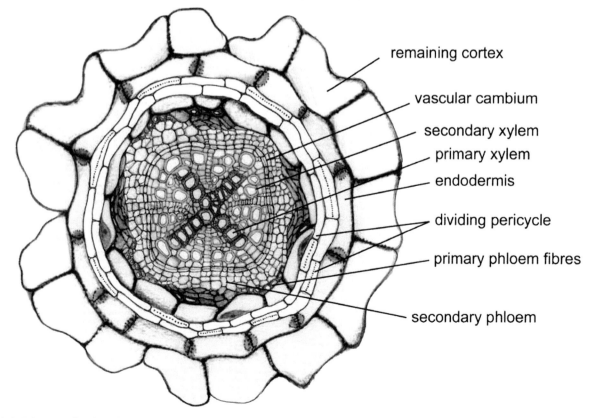

Figure 181b. Diagram of a dicotyledonous root in early stage of secondary growth.

Box 18. Clearing and staining for lateral root primordia without sectioning

Method

1. Excise roots from plants and fix for a minimum of 24 hours in 50% ethanol, then rinse thoroughly with water.
2. Excise pieces of root just above and below where the first lateral roots are visible and leave overnight in 5% chromic acid at room temperature. Rinse well with water.
3. Stain 20 minutes in acetocarmine **(see Appendix 2)** at 60 °C. Place an iron needle in the staining vessel that adds a small amount of iron to enchance the staining. Rinse well.
4. Pass through an ethanol series (50%, 70%, 95%) to absolute ethanol, then 1:1 ethanol : methyl benzoate, then methyl benzoate alone.
5. View with a stereo binocular microscope or compound microscope. Use caution with methyl benzoate fumes. N.B. For tree roots, repeat step 2 with fresh chromic acid if clearing is not completed overnight. Brownish phenol or tannin deposits may be removed by soaking roots in 3% hydrogen peroxide for 2–18 hours.

Results

Clearing will remove cytoplasm, rendering the roots transparent. Acetocarmine will stain the nuclear material in lateral root apical **meristems** red, making them easily visible.

Reference: Hackett, C., and Stewart, H.E. 1969. A method for determining the position and size of lateral primordia in the axes of roots without sectioning. Annals of Botany 33: 679–682.

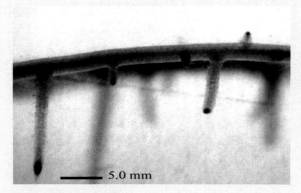

Pinus sp. lateral roots.
Photos courtesy of Daryl Enstone

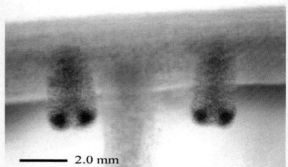

Pinus sp. dichotomous lateral roots.

Specialized roots

Aerial roots of orchids

Epiphytic orchids develop **aerial roots** that have many unusual features. Obtain an orchid species with aerial roots, prepare thin transverse sections of a root, and stain them in Toluidine Blue O. A distinctive feature of these roots is the development of a multi-layered epidermis (**velamen**) that consists of dead cells, often with lignified walls. These roots also have a pronounced exodermis, a cortex (with chloroplasts), an endodermis, a vascular cylinder, including a pith (Fig. 182). Cells in the velamen have secondary cell wall deposits that are lignified (Figs. 183, 184). Bands of lignified secondary cell walls are present in the cortex (Fig. 185). The outer row of cortical cells is differentiated as an exodermis (Fig. 186), in which modified passage cells called **tylosomes** occur (Fig. 187). Most of the cells in the exodermis (Figs. 186, 187) and endodermis (Fig. 188) develop lignified secondary walls. Passage cells are evident in the endodermis in Fig. 188.

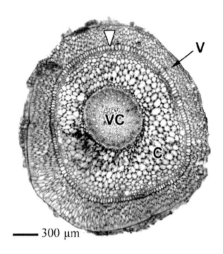

Figure 182. Transverse section of an aerial root of the orchid *Brassavola nodosa* stained with TBO showing the velamen (V), exodermis (arrowhead), cortex (C), and vascular cylinder (VC).

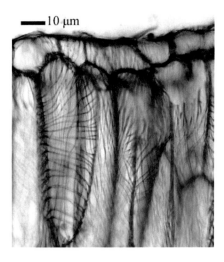

Figure 183. Transverse section of an aerial root of the orchid *Brassavola nodosa* stained with TBO showing variation in the lignified wall thickenings of velamen cells.

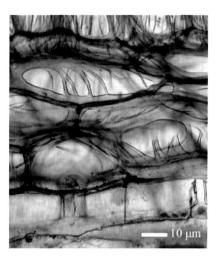

Figure 184. Transverse section of an aerial root of the orchid *Brassavola nodosa* stained with TBO showing variation in the lignified wall thickenings of velamen cells.

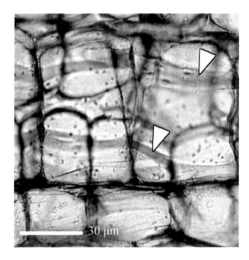

Figure 185. Transverse section of an aerial root of the orchid *Cattleya* sp. stained with TBO showing lignified bands of secondary cell wall (arrowheads) in cortical cells.

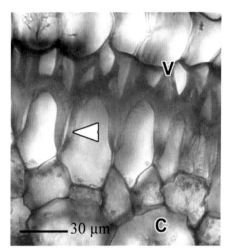

Figure 186. Transverse section of an aerial root of the orchid *Cattleya aurantiaca* stained with TBO showing a portion of the two–cell-layered velamen (V) and the exodermis that has developed lignified secondary cell walls (arrowhead). Cells in the cortex (C) contain chloroplasts.

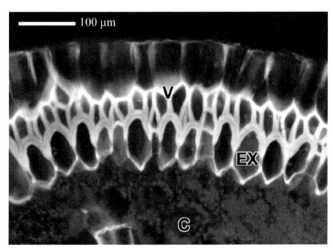

Figure 187. Unstained transverse section of an aerial root of the orchid *Cattleya aurantiaca* viewed with UV light. Cell walls in the velamen (V), especially the innermost layers of cells, and exodermis (EX) show autofluorescence due to the presence of lignin. Chloroplasts in the cortex (C) are also autofluorescent, appearing red when viewed with UV light.

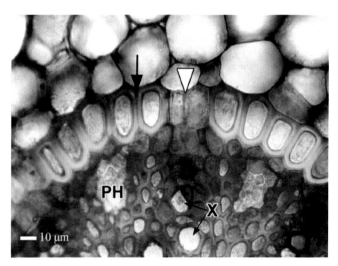

Figure 188. Transverse section of an aerial root of the orchid *Phalaenopsis* sp. stained with TBO showing endodermal cells with thick, lignified secondary cell walls (arrow). Two passage cells (arrowhead) are present. PH = phloem; X = xylem tracheary elements.

Storage roots

Roots of many species become highly modified as storage organs. In most cases these roots undergo a considerable amount of secondary growth of an unusual nature. In all cases, an abundance of parenchyma differentiates within secondary tissues, and various storage compounds are deposited within these parenchyma cells.

Carrot (*Daucus carota*)

Carrot roots undergo considerable secondary growth, with the bulk of the root consisting of parenchyma in the secondary xylem and secondary phloem (Fig. 189). In this case, there is more secondary phloem than secondary xylem, an unusual feature for roots. The main storage compound in carrot roots is sugar but these roots also have high levels of potassium, calcium, phosphorus, and some starch. Prepare thin sections of carrot roots (choose roots that are not too large or trim the root to achieve a small cutting face) and stain with Toluidine Blue O. Interpret the organization of tissues with the help of Figures 190 and 191.

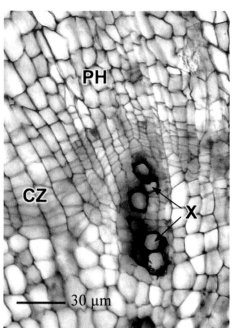

Figure 190. Montage of a transverse section of a carrot (*Daucus carota*) root stained with TBO. The large amount of secondary phloem (PH) consisting mainly of parenchyma for storage is evident. A vascular ray (arrowheads) is present in the secondary xylem (X) and the secondary phloem. The cambial zone (double arrowhead) is evident.

Figure 191. Transverse section of a carrot (*Daucus carota*) root stained with TBO showing the cambial zone (CZ) the derivatives of which give rise to the secondary phloem (PH) and secondary xylem (X).

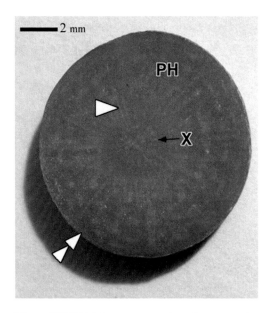

Figure 189. Thick transverse section of a carrot (*Daucus carota*) root showing the extent of secondary growth and the predominance of secondary phloem (PH) in comparison to secondary xylem (X). Vascular rays (arrowhead) are particularly evident in the secondary phloem. The outer tissue (double arrowhead) is a combination of pericycle and phellem.

Sweet potato (*Ipomoea batatas*)

These roots also have a proliferation of parenchyma cells but with starch as the main storage compound. They have a very unusual feature in the differentiation of additional vascular cambia surrounding individual vessels or vessel groups in the xylem. Prepare thin sections of a portion of a root, stain some with Toluidine Blue O and others with I_2KI. Identify vessels and the vascular cambia surrounding them (Fig. 192), the **compound starch grains,** and the druse crystals in parenchyma cells (Fig. 193).

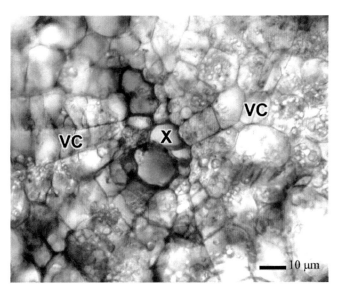

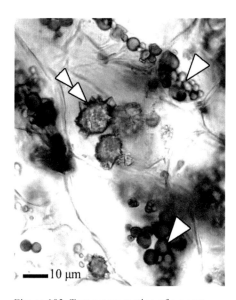

Figure 192. Transverse section of a sweet potato (*Ipomoea batatas*) root stained with TBO showing an anomalous vascular cambium zone (VC) surrounding a group of secondary xylem tracheary elements (X).

Figure 193. Transverse section of a sweet potato (*Ipomoea batatas*) root stained with I$_2$KI showing starch grains (arrowheads) and druse crystals (double arrowhead).

Roots of aquatic plants

The main variations in structure of these roots are the development of **aerenchyma** and the reduction in the amount of vascular tissues that form. Several aquatic plants can be used to demonstrate these features.

Water hyacinth (*Eichhornia crasipes*)

Prepare thin transverse sections of roots and stain with Toluidine Blue O. Locate the large air channels in the cortex (Fig. 194) comprising the aerenchyma, and note the reduction in phloem and xylem. Roots of this species develop numerous lateral roots.

Purple loosestrife (*Lythrum salicaria*)

This species can grow under inundated conditions as well as in drier habitats. Adventitious roots formed along submerged stems develop a unique **aerenchyma tissue** that is formed from derivatives of the **phellogen**. Prepare thin transverse sections of adventitious roots, stain with Toluidine Blue O, and locate the aerenchyma that consists of branched parenchyma cells and large air channels (Fig. 195).

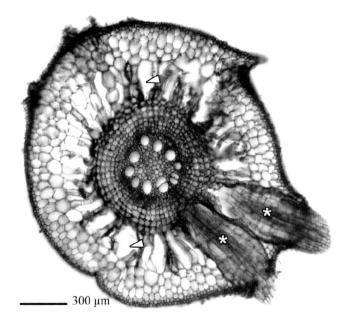

Figure 194. Transverse section of a water hyacinth (*Eichhornia crasipes*) root stained with TBO showing large air channels (arrowhead) in the cortex. Two lateral roots (∗) are evident.

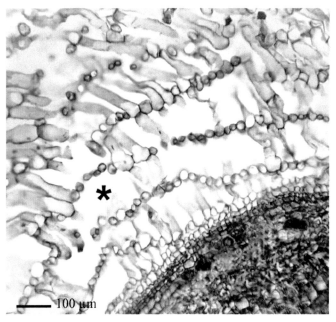

Figure 195. Transverse section of a purple loosestrife (*Lythrum salicaria*) root stained with TBO showing aerenchyma tissue consisting of large air channels (∗) and branched parenchyma cells. Photo courtesy of Kevin Stevens.

Chapter 8
Stems

Shoot apex

The shoot **apical meristem** and newly initiated **leaf primordia** can be demonstrated from living material because, unlike root apical meristems, these are not covered by a special tissue.

Coleus (*Coleus blumei*)

Any variety of coleus may be used. Using a seedling, cut the stem below some of the last-formed, expanded leaves and, using fine probes (sewing needles mounted in holders work well) and working under a binocular dissecting microscope, fold back the young leaves until they break off from the stem. Continue this until all that remains are leaf primordia. Carefully fold these back until the dome-shaped apical meristem is visible (Fig. 196). Leaf primordia of coleus are arranged in an opposite pattern; axillary buds and developing trichomes will be evident (Fig. 196).

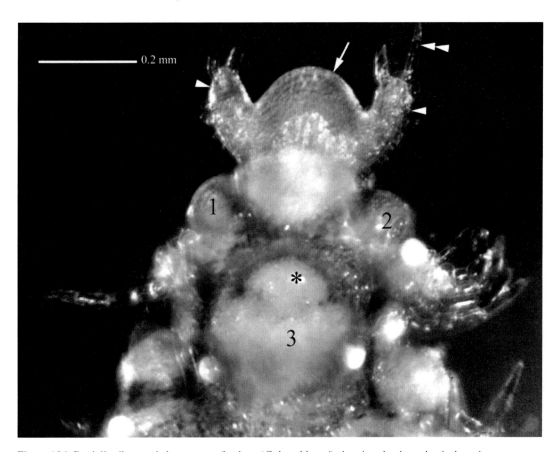

Figure 196. Partially dissected shoot apex of coleus (*Coleus blumei*) showing the domed apical meristem (arrow), leaf primordia (arrowheads), a developing axillary bud (∗) and trichomes (double arrowhead). Structures indicated by 1, 2, 3 are the bases of removed leaf primordia. Photo courtesy of Trevor Wilson.

Canadian pondweed (*Elodea canadensis*)

Young shoots of this species are very good specimens to illustrate the elongated apical meristem and numerous leaf primordia typical of many monocots. Using the same method as for coleus, remove the many young leaves that surround the apical meristem to demonstrate the elongated apical meristem and the ridge-like leaf primordia (Fig. 197).

Corn (*Zea mays*)

Shoot apices that have transitioned into flowering are particularly interesting to dissect because of their complexity. For example, the staminate inflorescence (tassle) meristem develops a series of spikelet primordia (Fig. 198) that give rise to the male flowers.

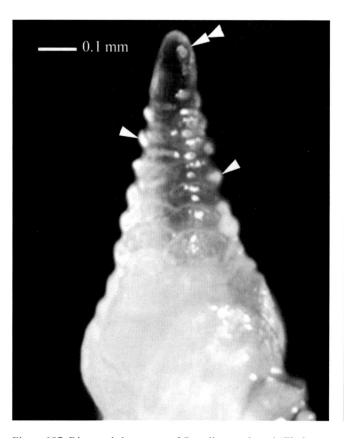

Figure 197. Dissected shoot apex of Canadian pondweed (*Elodea canadensis*) showing the elongated apical meristem (double arrowhead) and leaf primordia (arrowheads). Photo courtesy of Trevor Wilson.

Figure 198. Dissected apex of corn (*Zea mays*) showing the staminate inflorescence (tassle) meristem (arrow) and spikelet primordia (arrowheads). Photo courtesy of Ryan Geil.

Primary tissues in stems

There is tremendous variation in the organization of primary tissues in stems, and only a few of these are represented by the suggested species.

Sunflower (*Helianthus annuus*)

Prepare thin transverse sections several **internodes** back from the shoot apex and stain with Toluidine Blue O. The basic organization of primary tissues in the stem of this dicotyledonous species is illustrated in Fig. 199. When studying these stems, pay attention to the cell types present in the cortex and pith. The arrangement of the vascular bundles in a single ring is typical of many dicotyledonous species. Vascular bundles such as these have primary phloem and primary xylem arranged along the same radius (Fig. 200), unlike the situation in roots where the protoxylem and the phloem are on alternating radii. Vascular bundles in sunflower stems have one area of primary phloem and one area of primary xylem; these are called **collateral vascular bundles**.

Protoxylem is located towards the pith, and metaxylem is external to the protoxylem. The primary phloem has a cap of fibres (**bundle cap**) located next to the cortex (Fig. 200).

Tomato (*Solanum lycopersicum*)

Prepare thin sections of an internode several nodes below the shoot apex and stain with Toluidine Blue O. Tomato, like all members of the family Solanaceae, has **bicollateral vascular bundles**, with primary phloem differentiating on both sides of the primary xylem (Fig. 201). The cortex consists of one to two rows of cells with many chloroplasts, several rows of angular collenchyma cells, and several rows of large, thin-walled parenchyma cells. The vascular bundles are not as distinct as in sunflower because of the onset of secondary growth but protoxylem and metaxylem are still evident (Fig. 202). Sclerified cells are associated with both the internal and external phloem, which consists of sieve tube members, companion cells, and parenchyma cells.

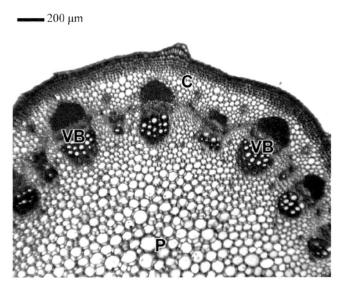

Figure 199. Transverse section of sunflower (*Helianthus annuus*) stem stained with TBO showing primary tissue organization. A ring of vascular bundles (VB) separates the cortex (C) from the pith (P).

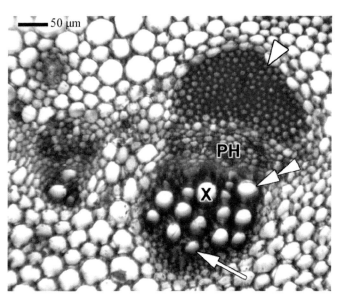

Figure 200. Transverse section of sunflower (*Helianthus annuus*) stem stained with TBO showing a vascular bundle with primary xylem (X) consisting of protoxylem (arrow), metaxylem (double arrowhead) and primary phloem (PH) arranged in a collateral fashion. A cap of phloem fibres (arrowhead) has developed.

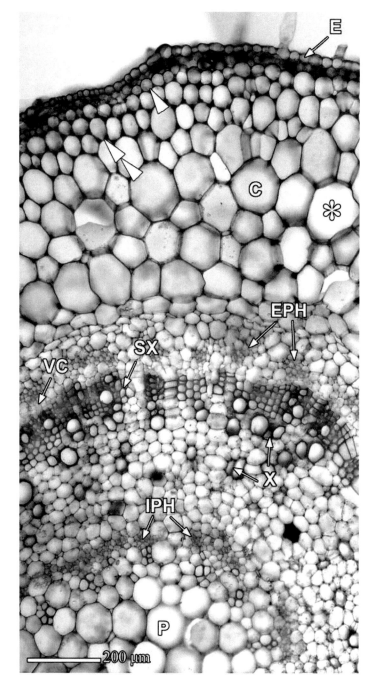
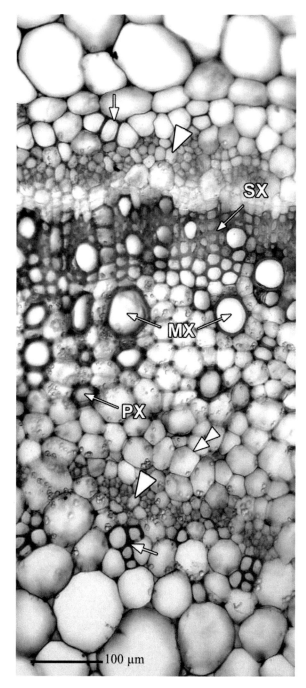

Figure 201. Transverse section of tomato (*Solanum lycopersicum*) stem stained with TBO showing epidermis (E), cortex (C), consisting of cells with chloroplasts (arrowhead), angular collenchyma (double arrowhead), and large parenchyma cells (∗). External phloem (EPH), vascular cambium (VC), secondary xylem (SX), primary xylem (X), internal phloem (IPH), and pith (P) are present.

Figure 202. Transverse section of tomato (*Solanum lycopersicum*) stem stained with TBO showing the bicollateral nature of the vascular tissue. The external and internal phloem both consist of sieve tube members (arrowheads), parenchyma cells (double arrowhead) and fibres (arrows). Secondary xylem (SX) and primary xylem consisting of metaxylem (MX) and protoxylem (PX) are present.

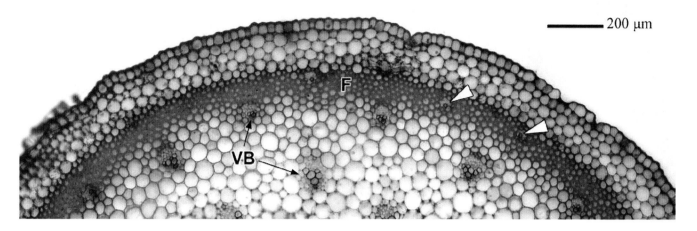

Figure 203. Transverse section of a spider plant (*Chlorophytum comosum*) runner (stem) stained with TBO. The larger vascular bundles (VB) are located in ground tissue (the term used when distinction between cortex and pith is not precise). A band of fibres (F) is present external to the larger vascular bundles. A few very small vascular bundles (arrowheads) are found within this region.

Spider plant (*Chlorophytum comosum*)

The runners of this species are ideal for illustrating the arrangement of vascular bundles in many monocotyledonous species. Prepare thin sections, stain with Toluidine Blue O, and observe the scattered arrangement of the vascular bundles (Fig. 203). Other features to note are the lamellar collenchyma in the epidermis and the band of lignified fibres in the cortical region. At higher magnification the collateral arrangement of primary xylem and primary phloem is evident (Fig. 204). Vascular bundles in stems of monocotyledonous species are referred to as **closed vascular bundles** because they do not contain precursors of vascular cambium and, therefore, the stems do not undergo secondary growth.

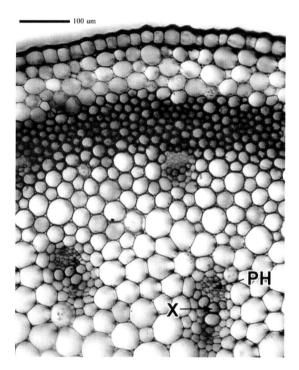

Figure 204. Transverse section of a spider plant (*Chlorophytum comosum*) runner (stem) stained with TBO showing the collateral arrangement of primary phloem (PH) and primary xylem (X) in each vascular bundle.

Corn (*Zea mays*)

Seedlings with several extended internodes or more mature plants should be selected. Prepare thin transverse sections of mature regions of this monocotyledonous species and stain with Toluidine Blue O. Vascular bundles surrounded by **sclerified bundle sheaths** are randomly distributed in ground tissue (Fig. 205). Each collateral vascular bundle consists of primary phloem and primary xylem organized in a collateral pattern (Figs. 205, 206). The sieve tube members and accompanying companion cells are very pronounced in the phloem as are the **protoxylem lacuna** and metaxylem in the xylem (Fig. 206). A cap of fibres is present, external to the phloem. Sections stained with phloroglucinol-HCl show that the hypodermal cells and bundle sheath cells are lignified and provide support for the stem (Fig. 207).

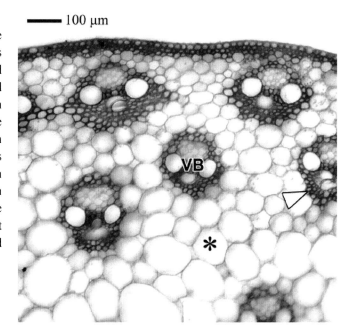

Figure 205. Transverse section of a corn (*Zea mays*) stem stained with TBO showing the random arrangement of vascular bundles (VB) in the ground tissue (∗). Each vascular bundle is surrounded by a bundle sheath of fibres (arrowhead).

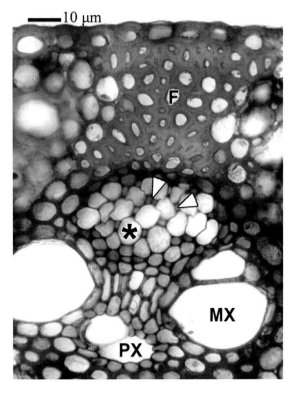

Figure 206. Transverse section of corn (*Zea mays*) stem stained with TBO showing the details of a vascular bundle. Primary phloem consists of sieve tube members (∗) and companion cells (arrowheads). Metaxylem (MX) and a protoxylem lacuna (PX) comprise the primary xylem. F = fibres.

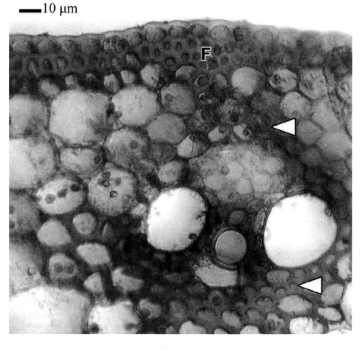

Figure 207. Transverse section of a corn (*Zea mays*) stem stained with phloroglucinol-HCl showing a band of fibres (F) beneath the epidermis and the bundle sheath of fibres (arrowheads).

Stems

There are numerous variations on the basic structure of stems and, as indicated in previous sections on cell types, stems are ideal to illustrate parenchyma, collenchyma, and sclerenchyma.

Beefsteak plant (*Iresine herbstii*)

Prepare thin transverse sections of *Irisene herbstii* (a dicotyledonous species) stems; mount some directly in water and stain others with Toluidine Blue O. Angular collenchyma cells with their whitish glistening thick walls as well as parenchyma cells containing anthocyanin pigments and compound starch grains are evident in unstained sections (Fig. 208). Stained sections show that cell walls of the angular collenchyma contain pectic substances, whereas fibres associated with the phloem, and both xylem tracheary elements and xylem parenchyma cells are lignified (Fig. 209).

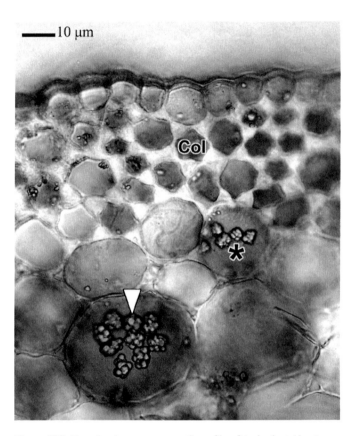

Figure 208. Unstained transverse section of beefsteak plant (*Iresine herbstii*) stem showing angular collenchyma (Col) and parenchyma cells in the cortex, some of which contain anthocyanin pigment (∗) and compound starch grains (arrowhead).

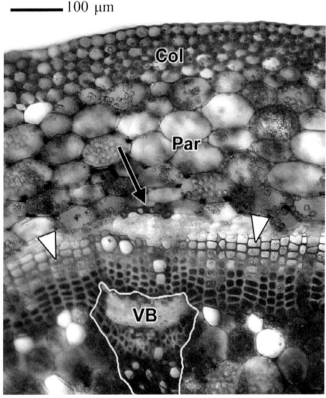

Figure 209. Transverse section of beefsteak plant (*Iresine herbstii*) stem stained with TBO showing angular collenchyma (Col), parenchyma cells (Par), sclerified parenchyma in the secondary xylem (arrowheads), and a vascular bundle, an unusual feature at this stage of development (outlined area marked with VB). Arrow indicates phloem fibres.

Wooly morning glory (*Argyreia nervosa*)

Stems of this species have an unusual arrangement of vascular bundles. Prepare thin transverse sections, stain with Toluidine Blue O, and observe the ring of primary xylem (part of the original vascular bundles) towards the cortex. This stem has gone into secondary growth so that the primary phloem of the original vascular bundles has been displaced outwards. Additional **medullary vascular bundles** occur in the pith. Numerous covering trichomes develop from the epidermis (Fig. 210). The medullary vascular bundles have a collateral arrangement of primary phloem and primary xylem (Fig. 211) as do the bundles towards the cortex.

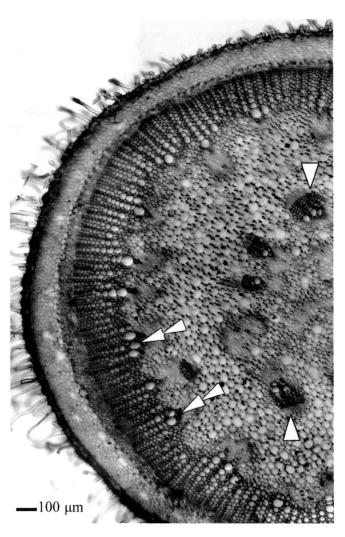

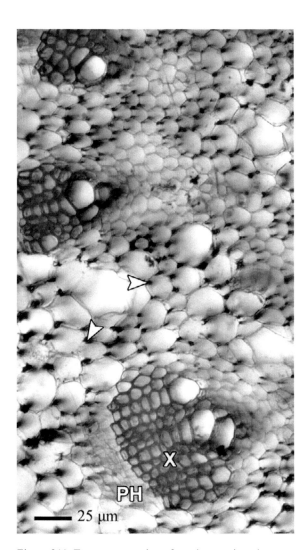

Figure 210. Transverse section of wooly morning glory (*Argyreia nervosa*) stem stained with TBO showing medullary vascular bundles (arrowheads) within the pith. The ring of vascular bundles (double arrowheads) are difficult to see because of the amount of secondary growth. Note the extensive development of trichomes from the epidermis, hence the name, wooly morning glory.

Figure 211. Transverse section of wooly morning glory (*Argyreia nervosa*) stem stained with TBO showing the collateral nature of a medullary vascular bundle. PH = phloem; X = xylem. Arrowheads indicate air trapped in intercellular spaces.

Secondary tissues in stems

The concept of secondary growth is often difficult for students to comprehend. Figure 212 shows, diagrammatically, the difference between primary growth and secondary growth in a stem as seen in longitudinal section. Primary tissues develop from derivatives (daughter cells) of the apical meristem, whereas secondary xylem and secondary phloem originate from derivatives of a **lateral meristem**, the **vascular cambium**. This cambium arises in two places in the stem, i.e., from cells between the primary xylem and primary phloem (**fascicular vascular cambium**) and in the parenchyma between adjacent vascular bundles (**interfascicular vascular cambium**). This is illustrated in Figure 213. A complete cylinder of vascular cambium is eventually formed. Although the vascular cambium is a single tier of cells, it is difficult to determine this because the recent derivatives are the same size. The term **cambial zone** is used to denote the vascular cambium and its most recent derivatives.

In addition, a second lateral meristem, the **phellogen** (**cork cambium**) arises in the cortex (or occasionally in the epidermis) and forms derivatives that differentiate into **phelloderm** (**secondary cortex**) and **phellem** (**cork**) (Fig. 212). The phellogen, phelloderm, and phellem make up the **periderm**.

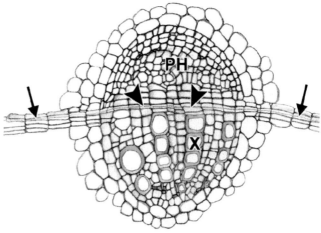

Figure 213. Origin of fascicular vascular cambium (arrowheads) and interfascicular vascular cambium (arrows) in transverse view. X = primary xylem; PH = primary phloem.

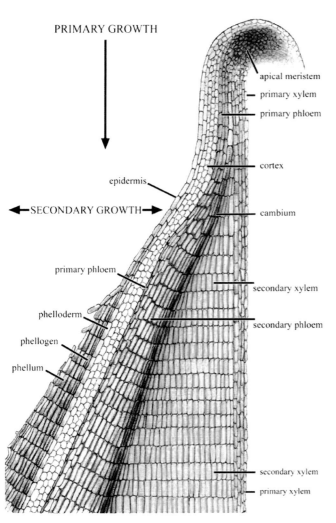

Figure 212. Diagram illustrating primary and secondary growth in a stem in longitudinal view.

Bean (*Phaseolus vulgaris*)

Prepare thin transverse sections of the stem at various distances from the shoot apex and stain with Toluidine Blue O. The vascular bundles in bean stems are widely spaced (Fig. 214) so that the initiation of the **interfascicular vascular cambium** as well as the **fascicular vascular cambium** is easy to detect (Fig. 215). With increasing distance from the apex, cambial activity becomes more pronounced and the formation of secondary xylem and secondary phloem becomes obvious (Figs. 216, 217).

Chrysanthemum (*Chrysanthemum morifolium*)

Prepare thin transverse sections from several internodes, beginning close to the shoot apex and at intervals basal to this and stain in Toluidine Blue O. The initiation of the vascular cambium and the first derivatives forming **secondary xylem** and **secondary phloem** should be evident in internodes fairly close to the shoot apex (Fig. 218). Older internodes should show considerable development of secondary xylem and secondary phloem (Fig. 219).

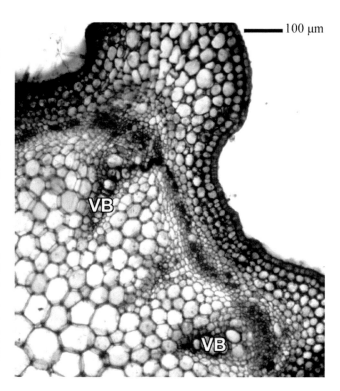

Figure 214. Transverse section of bean (*Phaseolus vulgaris)* stem stained with TBO showing the arrangement of vascular bundles (VB).

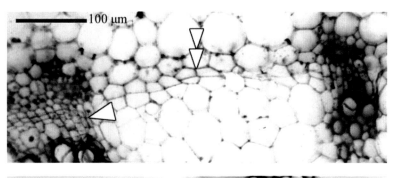

Figure 215. Transverse section of bean (*Phaseolus vulgaris*) stem stained with TBO showing the initiation of fascicular vascular cambium (arrowhead) and interfascicular vascular cambium (double arrowhead).

Figure 216. Transverse section of bean (*Phaseolus vulgaris*) stem stained with TBO showing the formation of secondary xylem (SX) from derivatives of the vascular cambium (arrowhead).

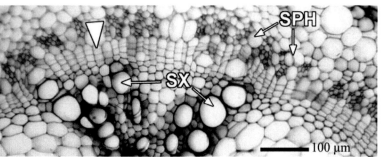

Figure 217. Transverse section of bean (*Phaseolus vulgaris*) stem stained with TBO showing the formation of secondary phloem (SPH) and secondary xylem (SX) from derivatives of the vascular cambium (arrowhead).

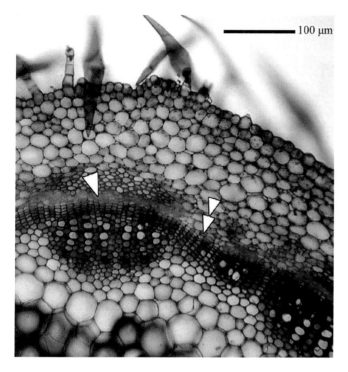

Figure 218. Transverse section of chrysanthemum (*Chrysanthemum morifolium*) stem stained with TBO showing the fascicular cambial zone (arrowhead) and the interfascicular cambial zone (double arrowhead).

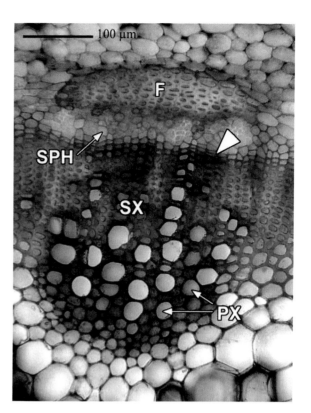

Figure 219. Transverse section of chrysanthemum (*Chrysanthemum morifolium*) stem stained with TBO showing the fascicular vascular cambium (arrowhead), secondary xylem (SX), and primary xylem (PX). Few secondary phloem elements (SPH) have been added. F = bundle cap of fibres that are a part of the primary phloem.

Periderm

In most stems that undergo secondary growth, the epidermis and any cortical cells external to the periderm are sloughed as a new protective tissue that prevents water loss and the ingress of pathogens is formed. The **phellogen (cork cambium)** is the lateral meristem responsible for forming new protective tissue. In most stems, the phellogen is initiated in the outer cortex (see Figure 212) but in some species it may arise in the epidermis or even the phloem.

Geranium (*Pelargonium hortorum*)

Prepare thin transverse sections of a stem internode that has a brown surface, indicating the formation of **phellem (cork)** by the lateral meristem, the **phellogen (cork cambium)**. In geranium, the phellogen originates in cells of the outer cortex and, by means of this meristem's **periclinal cell divisions**, several radially aligned rows of phellem and a single row of **phelloderm** are formed (Fig. 220). Walls of phellem cells contain suberin (a substance containing lipid and phenolic materials) that stains red–orange when stained with Sudan dyes (Fig. 220). In older portions of the stem, numerous rows of phellem cells are formed (Fig. 221). Mount some sections in water under a cover glass for viewing with UV or blue light using an epifluorescence microscope. Phellem cells show a yellow **primary fluorescence (autofluorescence)** when viewed with blue light (Fig. 222).

Potato (*Solanum tuberosum*)

Prepare thin sections of a potato tuber (a modified stem) making sure to include the 'skin' (periderm). Stain some sections with a Sudan dye and view with white light. Phellem cells will appear red–orange (Fig. 223). Mount other sections in water under a cover glass and view with UV light using an epifluorescence microscope. Suberin in phellem cell walls fluoresces a silvery-blue when viewed with UV light (Fig. 224).

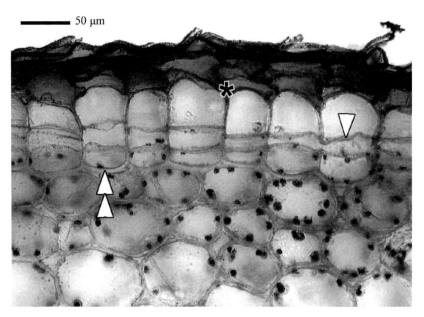
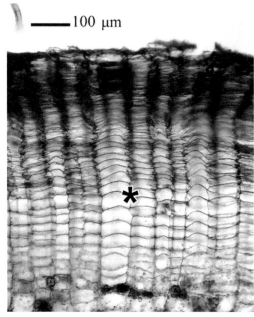

Figure 220. Transverse section of geranium (*Pelargonium hortorum*) stem stained with a Sudan dye showing phellogen (arrowhead), phelloderm (double arrowhead) and phellem (*). Walls of mature phellem cells stain orange–red because of the presence of suberin. Older phellem cells are collapsed.

Figure 221. Transverse section of geranium (*Pelargonium hortorum*) stem stained with a Sudan dye showing numerous layers of phellem cells (*).

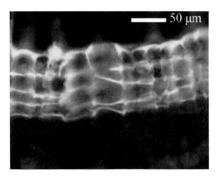
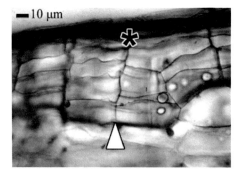
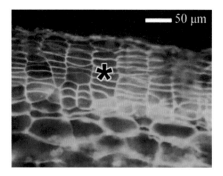

Figure 222. Unstained transverse section of geranium (*Pelargonium hortorum*) stem viewed with blue light. Phellem cells show autofluorescence because of the presence of suberin in their walls.

Figure 223. Transverse section of potato (*Solanum tuberosum*) tuber stained with a Sudan dye showing the phellogen (arrowhead) and suberized phellem cells (*).

Figure 224. Unstained transverse section of potato (*Solanum tuberosum*) tuber viewed with UV light. Phellem cells (*) show autofluorescence because of the presence of suberin in their walls.

Begonia (*Begonia rex*)

Cut thin transverse sections of the stem and stain some with Toluidine Blue O and others with phloroglucinol-HCl. The phellogen is initiated in the outer cortex and divides to form several rows of phellem cells (Fig. 225). **Lenticels**, loose regions of phellem cells that provide pathways for gas exchange (Fig. 226), often form in regions of the stem with a young periderm. The cortex of this stem consists of several rows of angular collenchyma (Fig. 225) and sclerified parenchyma cells (Figs. 225, 226).

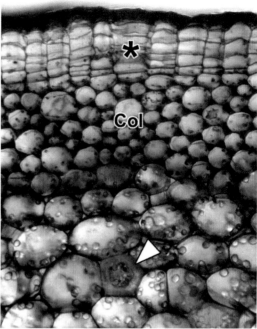

Figure 225.
Transverse section of begonia (*Begonia rex*) stem stained with TBO showing development of phellem (∗). The cortex (primary tissue) contains angular collenchyma (Col) and a sclerified parenchyma cell (arrowhead) in the cortex.

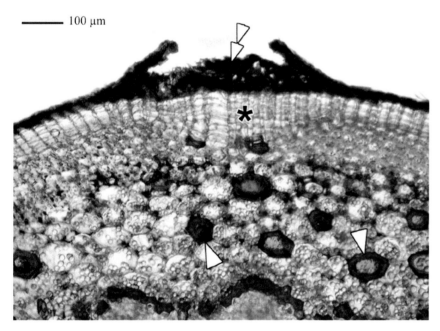

Figure 226.
Transverse section of begonia (*Begonia rex*) stem stained with phloroglucinol-HCl. A lenticel (double arrowhead) has developed in the outer phellem (∗). Several sclerified parenchyma cells that may develop into brachysclereids are present in the cortex (arrowheads).

Specialized stems

Stems of aquatic plants

As with roots of aquatic plants, the main structural modifications of stems of these plants is the development of **aerenchyma** and usually a reduction in the development of xylem and sometimes phloem. A number of species can be used to show these features, depending on those available.

Sedges such as dwarf spike rush (*Eleocharis parvula*), a monocotyledonous species, that grow in wet conditions show large air lacunae that extend from the cortex into the pith region (Fig. 227). Vascular bundles occur internal to the cortex that consists of radially elongated, photosynthetic cells (Figs. 227, 228). Groups of sclereids occur in the epidermis and provide support for the stem (Fig. 228). Each vascular bundle consists of phloem, a reduced amount of functional xylem, and a protoxylem lacuna (Fig. 228).

Submerged aquatics such as curly-leaved pondweed (*Potamogeton crispus*, Fig. 229) and water-milfoil (*Myriophyllum* sp., Fig. 230) also show well-developed aerenchyma consisting of large lacunae and reduced vascular tissue.

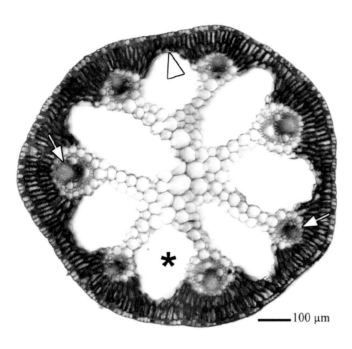

Figure 227. Transverse section of dwarf spike rush (*Eleocharis parvula*) stem stained with TBO. Large air lacunae (∗), a ring of vascular bundles (arrows), and photosynthetic parenchyma (arrowhead) are characteristic of this stem.

Figure 228. Transverse section of dwarf spike rush (*Eleocharis parvula*) stem stained with TBO showing a vascular bundle consisting of phloem (PH), xylem (X), and a protoxylem lacuna (∗). Bundles of sclereids (arrowhead) are located in the epidermis, and photosynthetic parenchyma (P) makes up the cortex.

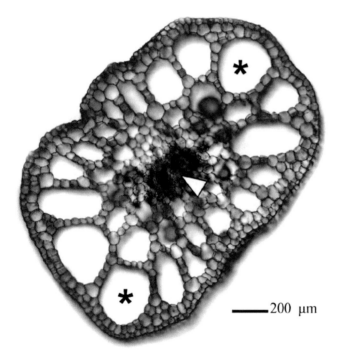

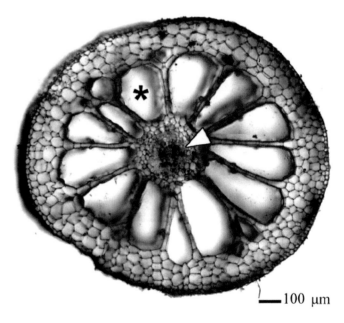

Figure 229. Transverse section of a curly-leaved pondweed (*Potamogeton crispus*) stem stained with TBO. Large air lacunae (∗) and a central region of vascular tissue (arrowhead) are evident.

Figure 230. Transverse section of a water-milfoil (*Myriophyllum* sp.) stem stained with TBO. Large air lacunae (∗) and a central region of vascular tissue (arrowhead) are present.

Stems with modified secondary growth

Dutchman's pipe (*Aristolochia durior*)

This dicotyledonous species shows an anomalous stem structure during secondary growth that can be observed by comparing transverse sections of internodes showing the beginning of secondary growth with those in which considerable secondary growth has occurred. Prepare thin transverse sections of internodes within two to three internodes of the shoot apex and of internodes below this, stain with Toluidine Blue O, and compare the organization of tissues between the two.

A ring of vascular bundles is present during primary growth, with each vascular bundle consisting of a collateral arrangement of phloem and xylem (Figs. 231, 232). A band of thick-walled, lignified fibres is present external to the vascular bundles. A fascicular and interfascicular vascular cambium develops (Fig. 232). With rapid cambial activity, considerable amounts of secondary xylem are produced. The thick ring of cortical fibres prevents outward growth and results in the pith becoming crushed by the inward growth of vascular tissues (Fig. 233). Eventually, the ring of cortical fibres is broken and is replaced by parenchyma of vascular rays (Fig. 233).

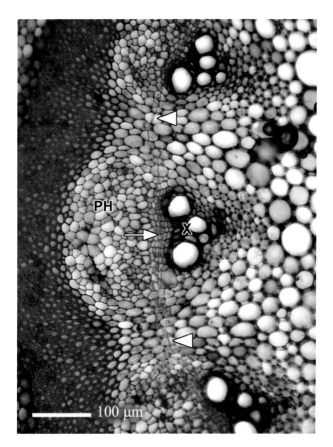

Figure 232. Enlargement of a portion of a Dutchman's pipe (*Aristolochia durior*) stem stained with TBO showing primary phloem (PH), primary xylem (X), fascicular vascular cambium (arrow), and interfascicular vascular cambium (arrowheads).

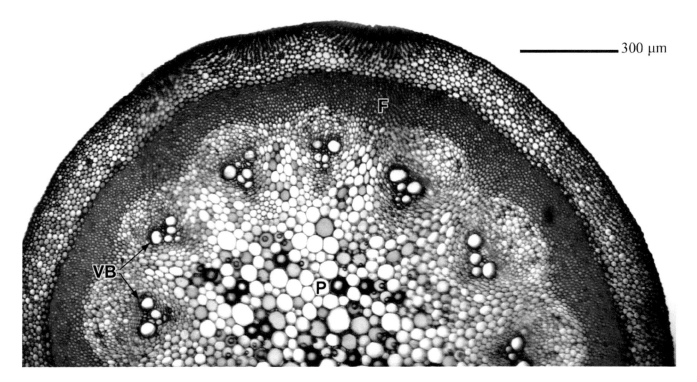

Figure 231. Transverse section of Dutchman's pipe (*Aristolochia durior*) stem in primary growth stained with TBO. A single ring of vascular bundles (VB), a complete band of fibres (F), and an extensive pith of thin-walled parenchyma cells (P) are present.

Trumpet vine (*Campsis radicans*)

This dicotyledonous species develops an additional vascular cambium in the pith, and this forms secondary xylem towards the existing primary xylem and secondary phloem towards the center of the stem. Prepare thin transverse sections at various distances from the shoot apex to study the initiation of the normal and additional vascular cambium and the products of each (Figs. 234, 235).

***Cordyline* spp.** and ***Dracaena* spp.** Species in these monocotyledonous genera can often be purchased as house plants. Stems develop a lateral meristem that forms derivatives in which **secondary vascular bundles** develop. Some species can undergo a considerable amount of lateral meristem activity leading to large-diameter stems. Prepare thin transverse sections of the stem and stain with Toluidine Blue O. The initial primary vascular bundles are evident towards the center of the stem, and the new secondary vascular bundles arising within the cortical tissue are evident in a more peripheral position (Fig. 236). Enlargements of a developing vascular bundle (Fig. 237) and a mature vascular bundle (Fig. 238) show the changes that occur with development.

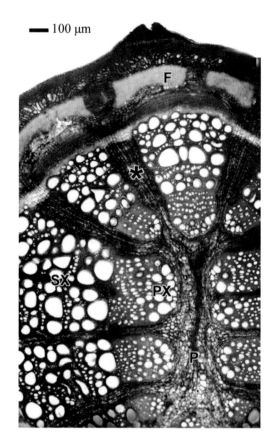

Figure 233. Transverse section of Dutchman's pipe (*Aristolochia durior*) stem in secondary growth stained with TBO showing the crushed pith (P), primary xylem (PX), secondary xylem (SX), large vascular rays (∗), and the broken band of fibres (F).

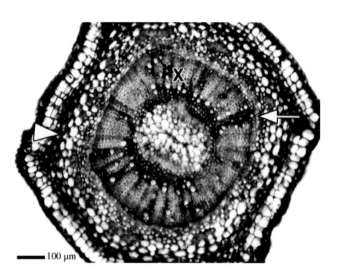

Figure 234. Transverse section of trumpet vine (*Campsis radicans*) stem stained with TBO. A solid ring of xylem (X), phloem (arrow), and a narrow cortex (arrowhead) are present.

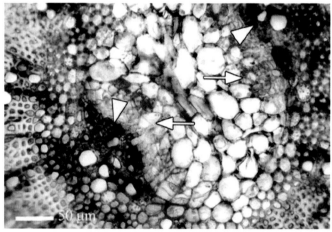

Figure 235. Transverse section of trumpet vine (*Campsis radicans*) stem stained with TBO. An enlargement of the pith region shows secondary phloem (arrows) and secondary xylem (arrowheads) derived from a cambium that was initiated from pith cells.

Stems

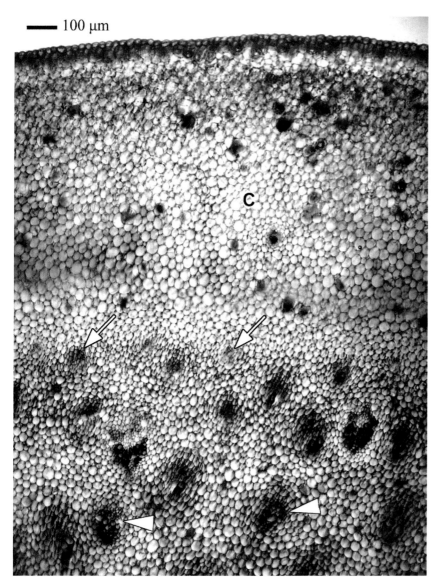

Figure 236.
Transverse section of *Dracaena* sp. stem stained with TBO. Low magnification showing the extensive cortex (C), the initiation of secondary vascular bundles (arrows), and mature vascular bundles (arrowheads).

Figure 237. Enlargement of a developing vascular bundle in the outer cortex of a transverse section of *Dracaena* sp. stem stained with TBO.

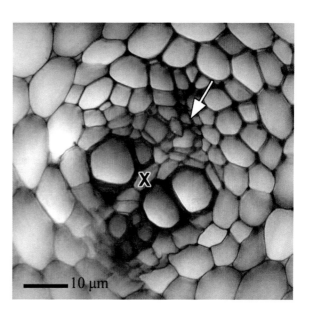

Figure 238. A mature vascular bundle in the inner cortex of a transverse section of *Dracaena* sp. stem stained with TBO showing phloem (arrow) and xylem (X).

Chapter 9
Leaves

Epidermis

Leaves of a number of species lend themselves to the examination of the organization of this primary tissue. Various methods can be used to characterize the cell types that develop in the epidermis.

Epidermal peels

Kalanchoe spp. and *Tradescantia* spp.

These species have leaves from which epidermal peels can be prepared **(see Box 19)**. Young stems often have an epidermis with some features in common with leaves. Epidermal peels from both surfaces of the leaf should be mounted in water and viewed unstained with white light. **Stomatal complexes**, consisting of **guard cells, subsidiary (accessory) cells** and the **stomatal pores (openings)** as well as ordinary epidermal cells, are evident in the lower epidermis (Figs. 239, 240) but stomatal complexes are not present in the upper epidermis of these species.

Wheat (*Triticum vulgare*)

Epidermal peels can be made from various grass leaves as well as cereals such as wheat and corn. The dumbbell-shaped, narrow guard cells and adjacent subsidiary cells are evident, especially when the tissue is stained with Toluidine Blue O (Fig. 241).

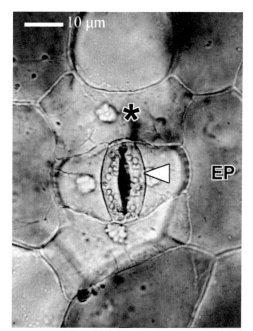

Figure 240. Epidermal strip of *Tradescantia pallida* leaf. Guard cells (arrowhead) are surrounded by four subsidiary cells (∗). Adjacent epidermal cells (EP) contain anthocyanin pigments.

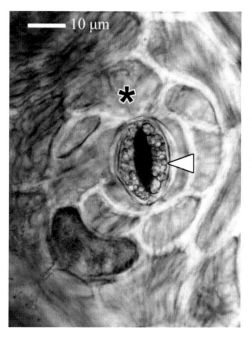

Figure 239. Epidermal strip of *Kalanchoe* sp. leaf. Guard cells with chloroplasts (arrowhead) are surrounded by several subsidiary cells (∗).

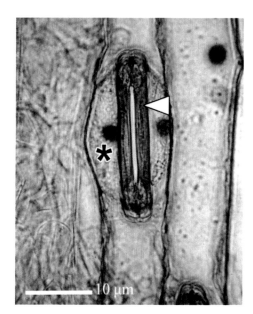

Figure 241. Epidermal strip of wheat (*Triticum vulgare*) leaf stained with TBO. The dumbbell-shaped guard cells (arrowhead) are associated with two subsidiary cells (∗).

Box 19. Method for making epidermal peels

Epidermal peels are usually made with leaves, although it is possible to remove small pieces of the epidermis from petals, stems, and even roots. Turn the leaf so that the epidermis you wish to examine is uppermost. Make a three-sided cut into the leaf (A) and fold the part enclosed by the cuts back onto the leaf (B). With forceps, pull gently on the flap (f) in the direction of the arrow and a piece of epidermis will often tear from the leaf (C). If the leaf is not too thick, the flap and its attached epidermal strip can be mounted directly onto a slide. However, if the leaf is too thick to mount, the flap should be removed (D) and only the epidermal peel (e) transferred to a slide.

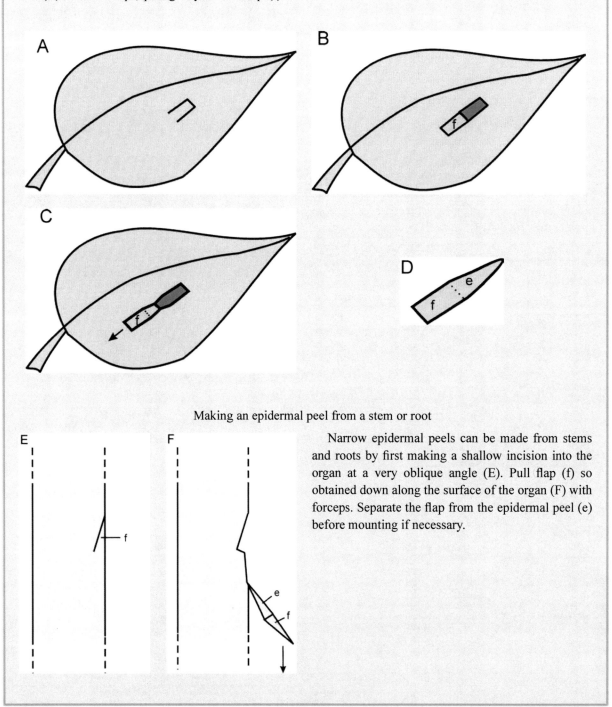

Making an epidermal peel from a stem or root

Narrow epidermal peels can be made from stems and roots by first making a shallow incision into the organ at a very oblique angle (E). Pull flap (f) so obtained down along the surface of the organ (F) with forceps. Separate the flap from the epidermal peel (e) before mounting if necessary.

Epidermal impressions

Leaves of species without too many trichomes can be used for this exercise. Clear nail polish is painted on both the upper and lower surfaces, allowed to dry, and then removed with fine forceps. Impressions of ordinary epidermal cells and stomatal complexes are evident (Fig. 242). Epidermal strips of the same species (Fig. 243) illustrate the accuracy of the information obtained by epidermal impressions.

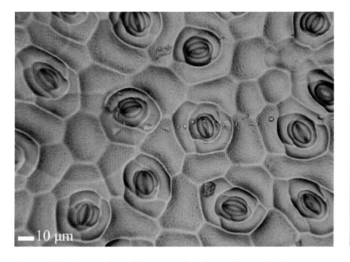

Figure 242. Impression of lower leaf surface of begonia (*Begonia rex*) leaf showing casts of numerous stomata.
Photo courtesy of Scott Liddycoat.

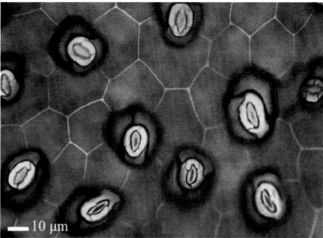

Figure 243. Epidermal strip of begonia (*Begonia rex*) leaf viewed with a combination of white and blue light. Numerous stomata are present. Photo courtesy of Scott Liddycoat.

Box 20. Peltate scales on bromeliad leaves

Most species of bromeliads (Bromeliaceae) are epiphytes growing on the limbs of trees, fence posts, wires, etc. Leaves of species in the subfamily Tillandsioideae have adapted to this habitat by developing specialized trichomes (peltate scales) that are able to trap and absorb water and minerals. Some species lack roots and, therefore, rely entirely on these scales for water and mineral absorption. Each trichome is multicellular consisting of a cap of dead cells and a stalk of living cells. The stalk is in intimate contact with mesophyll cells so that water and dissolved ions can be passed into the leaf. The peltate scales curl up under dry conditions and open when wet to allow water to be absorbed.

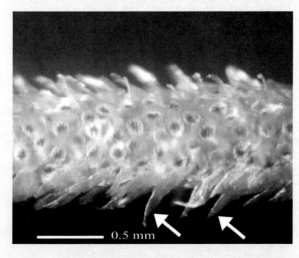

Portion of bromeliad leaf with numerous peltate scales (arrows).

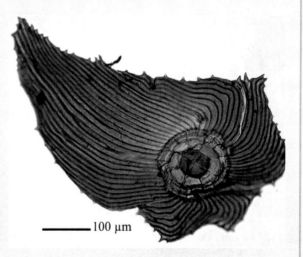

Peltate scale from bromeliad leaf stained with TBO.

Most leaves have a single layer of epidermal cells on their upper and lower surfaces but the size and contents of the ordinary epidermal cells can show much variation.

Prepare thin transverse sections of corn leaves using a Styrofoam support **(see Chapter 2)**, stain with Toluidine Blue O, and compare the structure of the upper and lower epidermis (Figs. 244, 245). Large cells in the upper epidermis are **bulliform cells** (Fig. 245); they lose their turgor during dry periods and cause leaf rolling.

Covering trichomes

The leaves of many species possess trichomes, usually with pointed terminal cells, that do not secrete compounds but serve other functions. These are called covering trichomes and they may function to deter herbivory by insects and other animals, prevent insects from ovi-depositing eggs on the surface of leaves, screen out some UV light, and reduce water loss by transpiration from the leaf. There are a multitude of trichome forms, varying from single celled to multicellular and from unbranched to highly branched. Some covering trichomes develop as **peltate scales (Box 20)**.

Plants can be screened by observing leaves with a stereobinocular microscope and a selection made from those with obvious covering trichomes. Some species will have secretory trichomes as well. Refer to Figures 114–118 for examples of these. Handle the leaves carefully so that the trichomes are not damaged. Prepare fairly thick sections of leaves with a sharp two-sided razor blade, mount in water under a cover glass, and observe the structure of the covering trichomes. Some examples of plant species and types of covering trichomes are illustrated in Figures 245–251.

Cleared leaves (**Boxes 21, 22**) are also useful to show the nature of covering trichomes (Figs. 252–254).

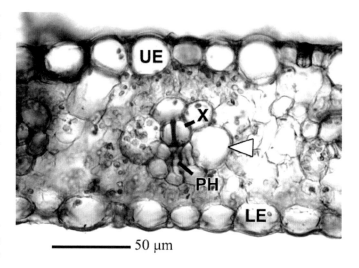

Figure 244. Transverse section of corn (*Zea mays*) leaf stained with TBO. The upper epidermis (UE) and the lower epidermis (LE) are evident. A conspicuous parenchyma bundle sheath (arrowhead) surrounds a **minor vein** consisting of a few tracheary elements in the xylem (X) and a small group of cells in the phloem (PH).

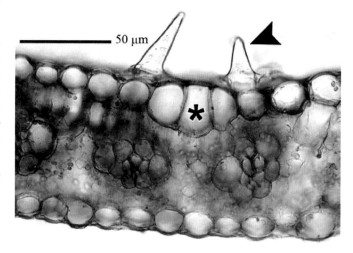

Figure 245. Transverse section of corn (*Zea mays*) leaf stained with TBO. Enlarged epidermal (bulliform) cells (∗) and small trichomes (arrowhead) are evident in the upper epidermis.

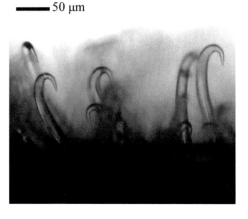

Figure 246. 'Hooked' covering trichomes on bean (*Phaseolus vulgaris*) leaf.

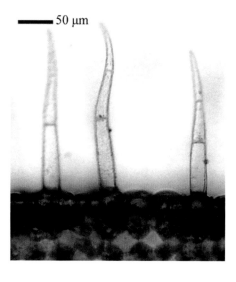

Figure 247. Covering trichomes on mother of thousands (*Saxifraga stolonifera*) leaf. Photo courtesy of Kirsten Otis.

Leaves

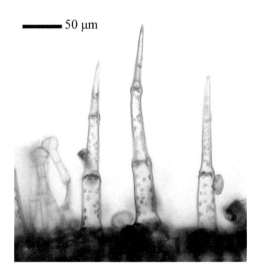

Figure 248. Covering trichomes on leaf of jimson weed (*Datura stramonium*).

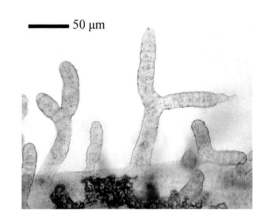

Figure 249. Branched covering trichomes on leaf of squirting cucumber (*Ecballium elaterium*).

Figure 250. Covering trichomes on leaf of chrysanthemum (*Chrysanthemum morifolium*). Photo courtesy of Ashleigh Downing.

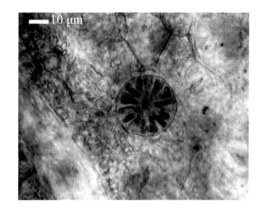

Figure 251. Top view of a peltate trichome on leaf of the aluminum plant (*Pilea cadierei*). Photo courtesy of Aidon Pyne.

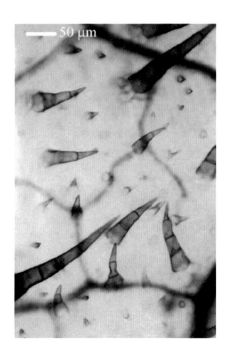

Figure 252. Covering trichomes on a cleared leaf of coleus (*Coleus blumei*) stained with safranin O.

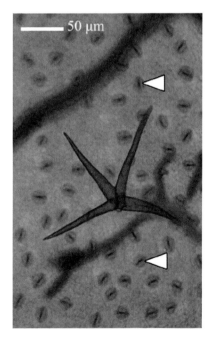

Figure 253. Branched covering trichome on a cleared leaf of ivy (*Hedera helix*) stained with safranin O. Many stomata (arrowheads) are also evident.

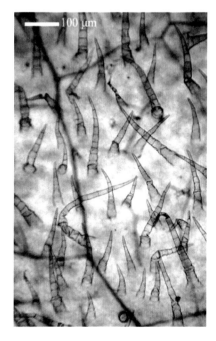

Figure 254. Covering trichomes on a cleared leaf of velvet plant (*Gynura sarmentosa*) stained with Chlorozol Black E. Photo courtesy of Greg Dillane.

Box 21. Clearing leaves

There are numerous clearing methods available to the plant anatomist. All have the same objective, i.e., to render the majority of the plant cells transparent by extracting most of their contents so that certain internal structures can be seen *in situ*. The choice of clearing method depends upon the pigmentation of the tissue, its thickness, and the structures one hopes to preserve. To kill tissue to facilitate extraction of the cell contents and to remove pigments, tissue is placed in acid alcohol (glacial acetic acid: 95% ethanol, 1:3). Cell contents are further solubilized by treatments with 2–5% NaOH or 85% lactic acid at temperatures up to 37 °C. To make a permanent preparation, the tissue is dehydrated in a series of alcohol concentrations, which clears the tissue further, and may then be mounted in a permanent mounting medium. Cell walls are not dissolved so that various cell patterns in tissues can be determined. For example, the number and arrangement of stomata on leaves can be ascertained from cleared leaves.

The most obvious structures remaining after most clearing procedures are the thick-walled lignified cells, since lignin is a very insoluble plant product. Lignin can also be stained with various dyes, such as safranin O to increase its visibility. This method is excellent for studying venation patterns, distribution of stomata, trichomes, and various cellular inclusions, particularly calcium oxalate crystals.

Reference: Gardner, R.O. 1975. An overview of botanical clearing techniques. Stain Technology 50: 99–105.

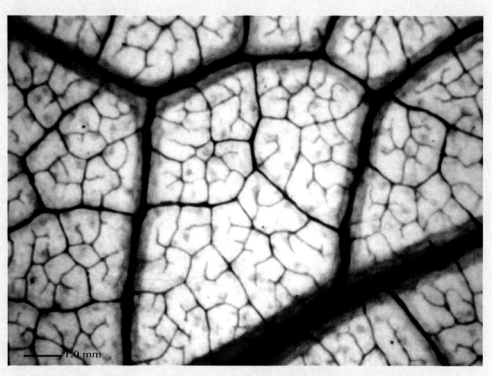

Cleared leaf of *Nasturtium* sp. stained with safranin O showing reticulate vein system.

Box 22. Simple clearing method for most leaves (and flat floral organs)

Leaves of many species, including those of *Arabidopsis thaliana,* are good specimens to clear using the following method:

1. Place leaves in a small, covered glass Petri dish in glacial acetic acid : 95% ethanol (1:3) and heat in an oven at 55 °C until most of the pigment has been extracted. If pigments are difficult to remove, decant solution and replace with fresh solution. Continue heating.
2. Place tissue in 85% lactic acid and heat at 55 °C for several hours (time depends on thickness of leaves). Leaves should become transparent.
3. Mount unstained tissue in lactic acid under a cover glass on a microscope slide. The cover glass can be ringed with nail polish to obtain a permanent slide.
4. To enhance the vasculature, stain with 0.03% chlorazol black E in lactoglycerol (85% lactic acid : glycerol 1:1) for 1 hour at 55 °C, de-stain for several hours in lactic acid, and mount in lactic acid under a cover glass on a microscope slide. Ring with nail polish to make a permanent slide.
5. Cleared tissue can be stored for months in lactic acid without mounting on slides.

For even faster results, see the method using commercial ink outlined in **Appendix 4**.

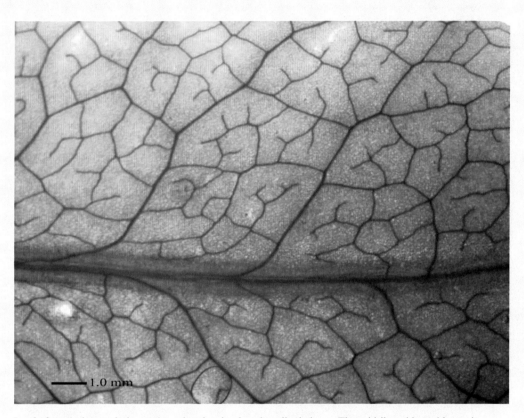

Leaf of *Arabidopsis thaliana* cleared and stained as described above. The midrib and branching vein system are evident.

Tissue organization in leaves

Leaves of most plants are challenging to section freehand but with practice good sections can be obtained, particularly if the leaf is supported by Styrofoam **(see Chapter 2)**.

Unstained transverse sections of leaves such as those of hibiscus (*Hibiscus* spp.) illustrate the common differentiation of the **mesophyll** into **palisade mesophyll**, usually located towards the upper epidermis, and **spongy mesophyll**, towards the lower epidermis (Fig. 255). Spongy mesophyll is so named because of the large amount of intercellular space in this tissue. Other arrangements occur such as having palisade mesophyll on both sides of the spongy mesophyll and, in the case of cylindrical leaves, having palisade mesophyll around the periphery of the leaf. Many grass species do not have distinct palisade and spongy mesophyll but have parenchyma cells of rather uniform size and shape. Most species have a uniseriate upper and lower epidermis (Fig. 255) but this can vary as well.

The fleshy leaves of the desert privet (*Peperomia magnoliaefolia*), for example, have a **multiple upper epidermis** consisting of large parenchyma cells that store water and provide a filter that protects the mesophyll from high-intensity light (Fig. 256). Transverse sections stained with Toluidine Blue O show this feature, the narrow row of palisade mesophyll, and a vascular bundle (vein) consisting of xylem towards the upper epidermis and phloem towards the lower epidermis (Fig. 256).

The thick, succulent leaves of the pearl plant or star window plant (*Haworthia* spp.) are easy to section. Prepare thin transverse sections and stain some with Toluidine Blue O and others with a Sudan dye. Stomata are sunken below the leaf surface and are over-arched by a cuticular ledge (Fig. 257). This arrangement reduces water loss from the leaf. These leaves also have a very thick, ridged cuticle as shown by staining with Sudan dye (Fig. 258).

Leaves (needles) of pine (*Pinus* spp.) also are easy to section, as shown previously for the study of resin ducts. Transverse sections stained with Toluidine Blue O show the undifferentiated mesophyll, resin ducts, the endodermis, **transfusion tissue**, and vascular bundles (Fig. 259). Figure 260 shows the sunken guard cells and the thick-walled, lignified epidermis and hypodermis.

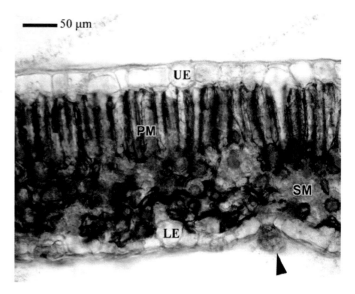

Figure 255. Unstained transverse section of *Hibiscus* sp. leaf showing upper (UE) and lower (LE) epidermis, palisade mesophyll (PM), spongy mesophyll (SM), and a secretory trichome (arrowhead).

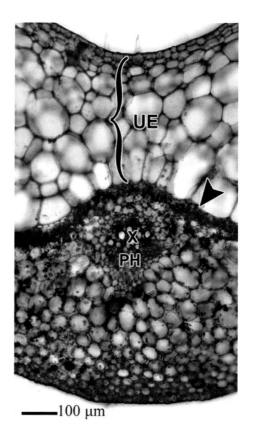

Figure 256. Transverse section of desert privet (*Peperomia magnoliaefolia*) leaf stained with TBO illustrating the multiple upper epidermis (UE), the narrow band of palisade mesophyll (arrowhead), and a vascular bundle consisting of xylem (X), and phloem (PH) surrounded by parenchyma.

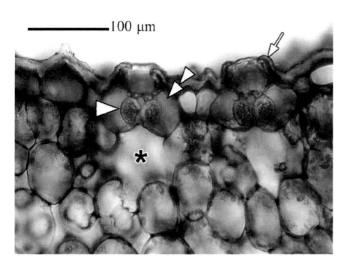

Figure 257. Transverse section of a pearl plant (*Haworthia* sp.) leaf stained with TBO showing guard cells (arrowhead), subsidiary cells (double arrowhead), **sub-stomatal chamber** (∗) and the overarching cuticular ledge (arrow).

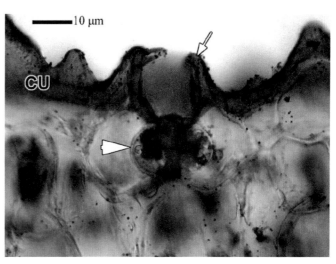

Figure 258. Transverse section of a pearl plant (*Haworthia* sp.) leaf stained with Sudan dye showing the thick cuticle (CU), cuticular ledge (arrow), and guard cells (arrowhead).

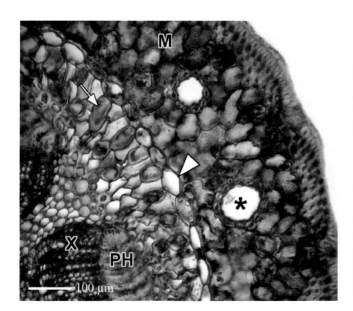

Figures 259. Transverse section of Scots pine (*Pinus sylvestris*) needle stained with TBO showing the undifferentiated mesophyll (M), resin ducts (∗), endodermis (arrowhead), transfusion tissue (arrow), phloem (PH), and xylem (X).

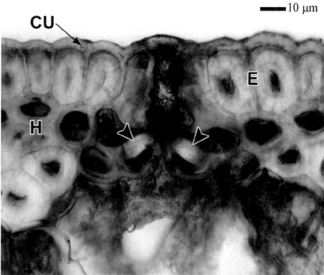

Figure 260. Transverse section of Scots pine (*Pinus sylvestris*) needle stained with TBO showing the cuticle (CU), thick-walled epidermis (E), hypodermis (H), and sunken guard cells (arrowheads).

Leaves of oleander (*Nerium oleander*) are instructive to show modifications of a dicotyledonous species that has become adapted to dry environments.

Prepare thin transverse sections and stain with Toluidine Blue O. Leaves have a multiple upper and lower epidermis, palisade mesophyll on both sides of the spongy mesophyll, and **stomatal crypts** (indentations in the lower epidermis) in which stomata are located (Figs. 261, 262). Each stomatal crypt contains a number of stomata and trichomes, the latter reducing water loss from the leaf (Fig. 262). The guard cells of the stomatal complex protrude into the crypt (Fig. 263). Large druse crystals occur in the mesophyll (Fig. 264). The **midvein** consists of a large amount of parenchyma in addition to phloem and xylem in a bicollateral arrangement (Fig. 265). Primary phloem occurs on both sides of the xylem (Fig. 266).

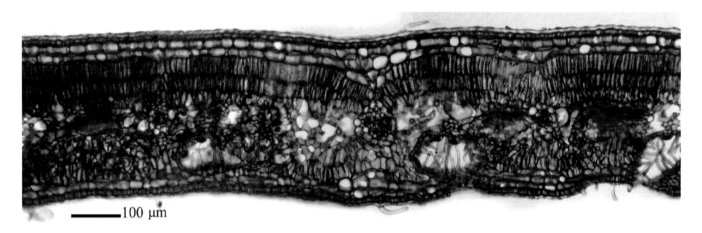

Figure 261. Transverse section of oleander (*Nerium oleander*) leaf stained with TBO showing overall tissue organization.

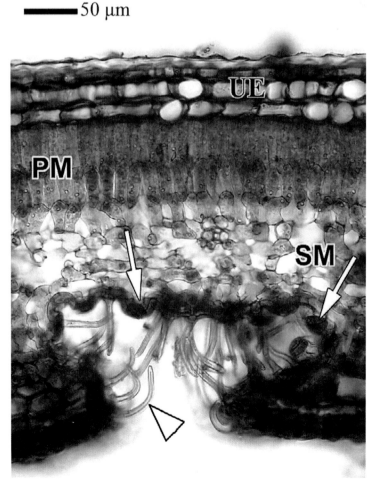

Figure 262. Transverse section of oleander (*Nerium oleander*) leaf stained with TBO showing multiple upper epidermis (UE), palisade mesophyll (PM), spongy mesophyll (SM), and a stomatal crypt with trichomes (arrowhead) and stomata (arrows).

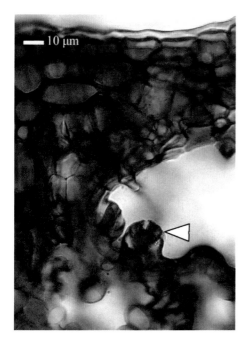

Figure 263. Portion of a stomatal crypt in a transverse section of oleander (*Nerium oleander*) leaf stained with TBO showing guard cells (arrowhead).

Figure 264. Transverse section of oleander (*Nerium oleander*) leaf stained with TBO showing a druse crystal (arrowhead) within a spongy mesophyll cell.

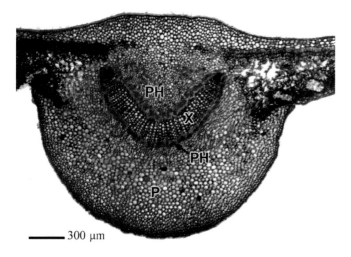

Figure 265. Midvein in an oleander (*Nerium oleander*) leaf stained with TBO with a large amount of parenchyma (P) forming the midrib, xylem (X), and phloem (PH).

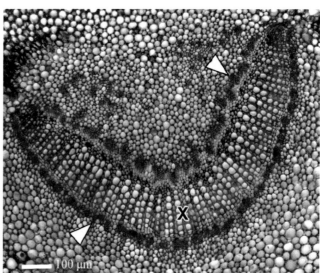

Figure 266. Midvein in an oleander (*Nerium oleander*) leaf stained with TBO with xylem (X) and phloem (arrowheads) on both sides of the xylem.

Monocot leaves

Prepare thin transverse sections of corn (*Zea mays*) leaves, stain some with phloroglucinol-HCl and others with Toluidine Blue O. If sections are prepared towards the base of a developing seedling, the arrangement of developing leaves is apparent (Fig. 267). Higher magnification of vascular bundles from more mature leaves shows the primary phloem, primary xylem with developing protoxylem lacunae, and lignified cells for support (Figs. 268, 269). The protoxylem lacunae result from the stretching and tearing of protoxylem tracheary elements during rapid elongation of the leaf.

The smaller bundles have few xylem and phloem cells but show distinct **bundle sheath cells** (Figs. 244, 245, 270). The arrangement of these cells around the vascular bundles is typical of many species with C_4 photosynthesis. The wreath-like arrangement of the photosynthetic parenchyma cells around the specialized bundle sheath cells (Fig. 270) is termed **Kranz anatomy**. Narrow guard cells with subsidiary cells, subtended by a **sub-stomatal chamber,** are evident in the epidermis (Fig. 271). Small unicellular trichomes and **bulliform cells** develop in the upper epidermis (Fig. 245).

Bundle sheath cells of C_4 plants often store large amounts of starch that can be shown dramatically by following the method outlined in **Box 23**. This same technique can be used to show starch deposition in bundle sheath cells of other C_4 plants compared to the uniform starch deposition in C_3 plants.

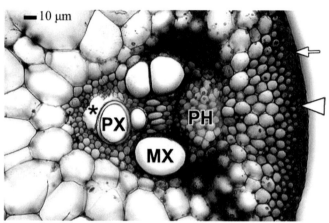

Figure 268. Enlargement of a vascular bundle from one of the leaves in Fig. 267 showing phloem (PH), metaxylem (MX), protoxylem (PX) and developing protoxylem lacuna (∗), and lignified epidermal (arrow) and hypodermal (arrowhead) cells.

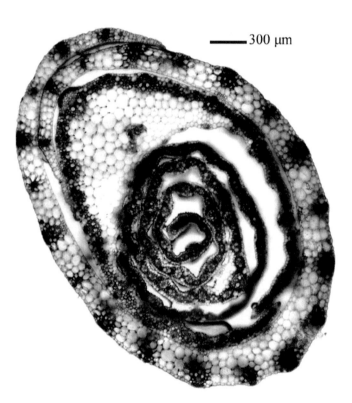

Figure 267. Transverse section through overlapping leaves at the base of a young corn (*Zea mays*) seedling stained with phloroglucinol-HCl.

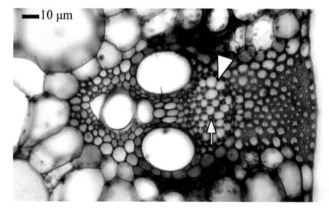

Figure 269. Transverse section through a corn (*Zea mays*) leaf stained with TBO. A vascular bundle showing similar features illustrated in Fig. 268. The phloem sieve tube members (arrowhead) and adjacent companion cells (arrow) are shown in more detail.

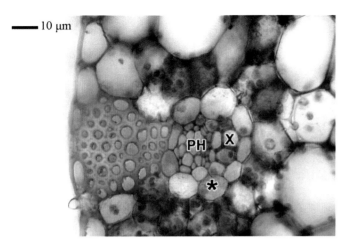

Figure 270. Transverse section through a corn (*Zea mays*) leaf stained with TBO. A small minor vein showing a small amount of phloem (PH) and xylem (X) and a bundle sheath (∗) containing large chloroplasts.

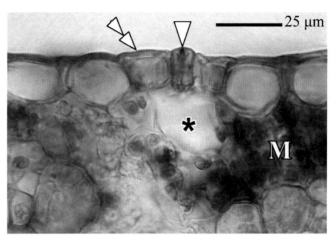

Figure 271. Transverse section of a corn (*Zea mays*) leaf stained with TBO. Guard cells (arrowhead), subsidiary cells (double arrowhead), a sub-stomatal chamber (∗), and mesophyll (M).

Prepare thin transverse sections of a spider plant (*Chlorophytum comosum*) leaf and mount in water, unstained. The mesophyll, containing chloroplasts is undifferentiated and the minor veins contain xlyem and phloem (Fig. 272).

Both upper and lower epidermal cells are covered by a distinct cuticle (Fig. 272). The sunken guard cells contain chloroplasts (Fig. 273).

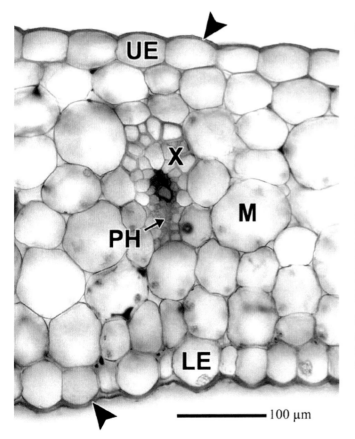

Figure 272. Transverse section of a spider plant (*Chlorophytum comosum*) leaf viewed unstained. The upper epidermal cells (UE) and lower epidermal cells (LE) are covered with a cuticle (arrowheads). The mesophyll (M) is undifferentiated and each vascular bundle consists of xylem (X) and phloem (PH).

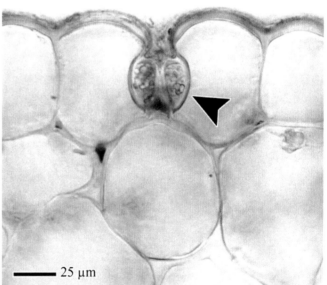

Figure 273. Transverse section of a spider plant (*Chlorophytum comosum*) leaf viewed unstained showing sunken guard cells (arrowhead) with chloroplasts. Adjacent epidermal cells lack chloroplasts.

Box 23. Demonstrating starch in bundle sheath cells of C_4 plants

Unlike leaves of C_3 plants that store starch in both mesophyll and bundle sheath cells, C_4 plants store starch predominantly in their bundle sheaths. In both cases, the location of this starch coincides with that of the Calvin cycle. The bundle sheaths enclose the veins of leaves. Thus for a C_4 plant, when a cleared leaf has been stained for starch, ones sees a wide-diameter outline of the vein system. In dicotyledonous species such as pigweed (*Amaranthus* sp.), a reticulate pattern (a) is evident, whereas in a monocotyledonous species such as corn (*Zea mays*), a parallel pattern is evident (b, c).

Method

1. Cut pieces of pigweed (*Amaranthus* sp.), corn (*Zea mays*), or other suitable leaves and place them in a beaker with 80% ethanol.
2. Heat gently (low heat) on a hotplate for ~30 minutes or until most pigments are extracted; replace ethanol if necessary.
3. Transfer leaf tissue to 85% lactic acid and heat gently in a fume hood for 20–30 minutes or leave at room temperature overnight. Tissue can be stored in lactic acid for several months.
4. Transfer tissue directly to freshly made I_2KI (0.2% I in 2% aqueous KI) for 30 seconds at room temperature. Extend time with age of stain.
5. Mount in 50% glycerol. Starch will stain blue to black.

Reference: Crookston, R.K., and Moss, D.N. 1973. Plant Physiology 52: 397–402.

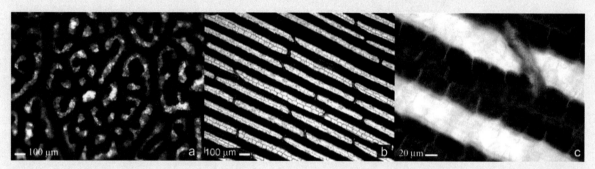

Photos courtesy of Ryan Geil and Chris J. Meyer

Vascular tissues (veins)

The vascular tissues (xylem and phloem) of leaves are generally referred to as **veins** and the arrangement of the veins as **venation**. There are various venation patterns, ranging from rather simple to very complex. The best way to observe these patterns is to clear leaves so that an overview of the pattern can be observed. A discussion of clearing methods for leaves is presented in **Boxes 21 and 22.**

Cleared leaves of crown of thorns (*Euphorbia splendens*) stained with safranin O show the typical **reticulate venation pattern** and illustrate **open venation** in a dicotyledonous species (Fig. 274). At higher magnification, the **minor veins** (Fig. 275) and **vein endings** with enlarged tracheary elements (Fig. 276) are apparent.

Cleared leaves of corn (*Zea mays*) show the **parallel venation pattern** with small cross veins (Fig. 277), typical of grass species. The large bundle sheath cells surrounding veins are also evident at higher magnification (Fig. 278).

Leaves of basswood (*Tilia americana*) have reticulate venation and illustrate **closed venation**. Numerous prismatic crystals are easily seen in this cleared leaf (Fig. 279).

Leaves cleared and stained by the method outlined in **Box 22** are illustrated in Figures 280–282. *Arabidopsis thaliana* leaves show a reticulate venation pattern and branched trichomes (Fig. 280). Leaves of geranium (*Pelargonium hortorum*) have numerous druse crystals in the mesophyll (Fig. 281), while leaves of chrysanthemum (*Chrysanthemum morifolium*) have reticulate venation and branched trichomes (Fig. 282).

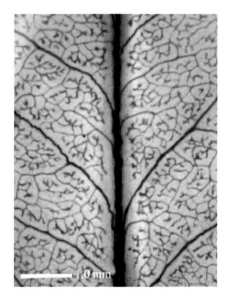
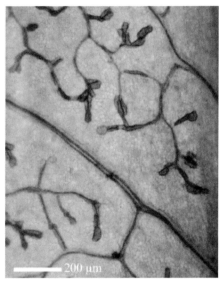
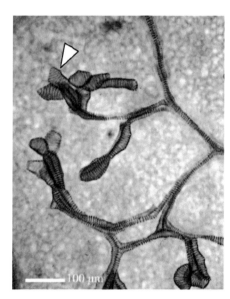

Figure 274. Cleared leaf of crown of thorns (*Euphorbia splendens*) stained with safranin O showing reticulate venation.

Figure 275. Cleared leaf of crown of thorns (*Euphorbia splendens*) stained with safranin O showing details of minor veins.

Figure 276. Cleared leaf of crown of thorns (*Euphorbia splendens*) stained with safranin O. Vein endings (arrowhead) consist of enlarged tracheary elements.

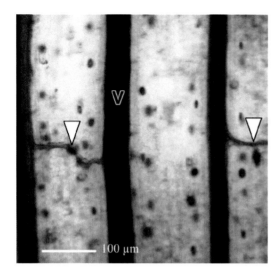

Figure 277. Cleared leaf of corn (*Zea mays*) stained with safranin O. Parallel veins (V) and small cross veins (arrowheads) are evident.

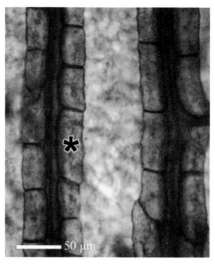

Figure 278. Cleared leaf of corn (*Zea mays*) stained with safranin O. Large bundle sheath cells (∗) surround the veins.

Figure 279. Cleared leaf of basswood (*Tilia americana*) stained with safranin O. Numerous prismatic crystals (arrowheads), many associated with the veins, are present.

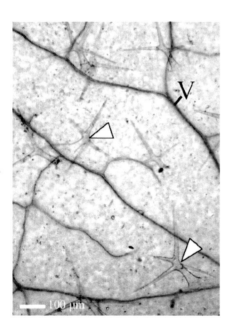

Figure 280. Cleared leaf of *Arabidopsis thaliana* stained with chlorozol black E showing veins (V) and branched trichomes (arrowheads).

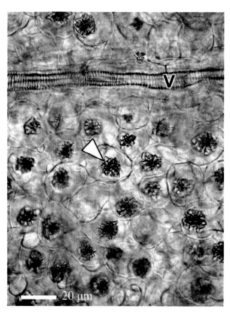

Figure 281. Cleared leaf of geranium (*Pelargonium hortorum*) stained with chlorozol black E showing a vein (V) and numerous druse crystals (arrowhead). Photo courtesy of Becky Longland.

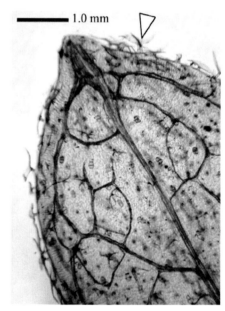

Figure 282. Cleared leaf of chrysanthemum (*Chrysanthemum morifolium*) stained with chlorozol black E. Reticulate venation and branched trichomes (arrowhead) are evident. Photo courtesy of Ashleigh Downing.

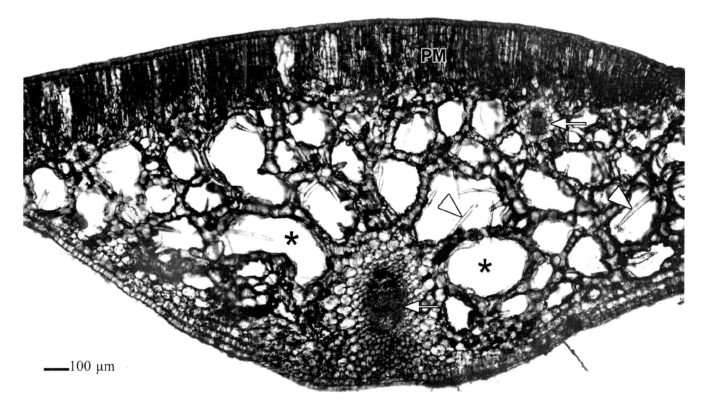

Figure 283. Transverse section of a water lily (*Nymphaea odorata*) leaf stained with TBO. Well-developed palisade mesophyll (PM), spongy mesophyll containing large air lacunae (∗) with arms of branched sclereids (arrowheads) protruding into them, and veins (arrows) are evident.

Leaves of aquatic plants

Leaves such as those of water lily (*Nymphaea odorata*) that float on the surface of water are fairly easy to section. Excise a strip from the broad leaf and trim it to a size that can be sectioned. Prepare transverse sections and stain with Toluidine Blue O. The arrangement of palisade mesophyll, spongy mesophyll containing large air lacunae, and veins is evident (Fig. 283). Arms of trichosclereids protrude into the air lacunae (Figs. 283, 284). Thick, unstained sections of petioles of the same species show the development of very large air lacunae and arms of trichosclereids (Fig. 285). The xylem within leaf blades and petioles (Fig. 286) consists only of a very large protoxylem lacuna.

The small leaves of water milfoil (*Myriophyllum* sp.) can be sectioned when held between pieces of Styrofoam **(see Chapter 2)**. Transverse sections stained with Toluidine Blue O show the very simple anatomy of these submerged structures (Fig. 287). A small group of phloem elements and a large xylem lacuna are surrounded by parenchyma cells and an epidermis.

Transverse sections of the submerged leaves of curly-leaved pondweed (*Potamogeton crispus*) show reduced vascular tissue and large air lacunae (Fig. 288).

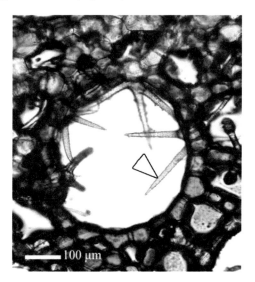

Figure 284. Air lacuna with arms of branched sclereids (arrowhead) in a transverse section of a water lily (*Nymphaea odorata*) leaf stained with TBO.

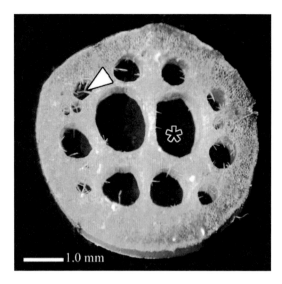

Figure 285. A thick transverse section of a petiole of water lily (*Nymphaea odorata*) viewed with a stereo binocular microscope showing large air lacunae (∗), some with arms of branched sclereids (arrowhead).

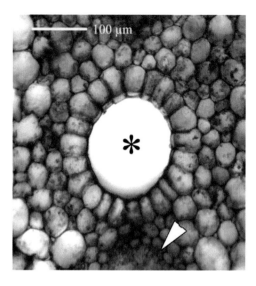

Figure 286. Phloem (arrowhead) and a xylem lacuna (∗) in a portion of a petiole of water lily (*Nymphaea odorata*) cross sectioned and stained with TBO.

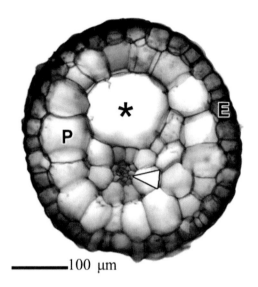

Figure 287. Transverse section of a leaf of water milfoil (*Myriophyllum* sp.) stained with TBO. Phloem (arrowhead) and a xylem lacuna (∗) are surrounded by parenchyma (P) and an epidermis (E).

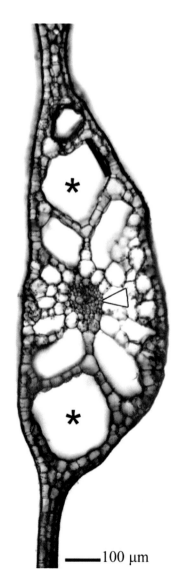

Figure 288. Transverse section of a leaf of curly-leaved pondweed (*Potamogeton crispus*) stained with TBO. This flattened, submerged leaf has reduced vascular tissue (arrowhead) and large air lacunae (∗).

Leaves

Petiole anatomy

Petioles of leaves contain the same basic tissues as stems. The vascular tissues, however, show a variety of arrangements that can be demonstrated in transverse sections stained with Toluidine Blue O. Some of the variation in vascular tissue arrangement is illustrated in Figures 289–296.

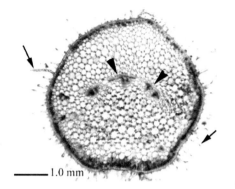

Figure 289. African violet (*Saintpaulia ionantha*). The vascular tissues are arranged in an arc (arrowheads) in the centre of the petiole. Numerous trichomes (arrows) are present.

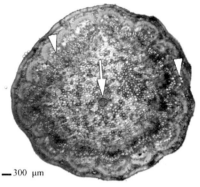

Figure 290. Horse-chestnut (*Aesculus hippocastanum*) petiole. A ring of vascular tissues (arrowheads) and a single small central vascular bundle (arrow) are present.

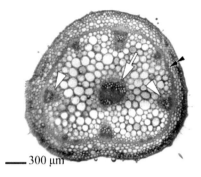

Figure 291. Geranium (*Pelargonium hortorum*) petiole. A ring of vascular bundles (arrowheads) and a large central vascular bundle (arrow) are present. A band of lignified cells (double arrowhead) is present in the cortex.

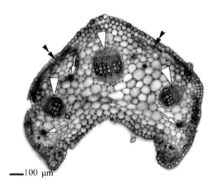

Figure 292. Marigold (*Tagetes erecta*) petiole. Three large (arrowheads) and a number of small (arrows) vascular bundles are present. Bands of parenchyma cells with chloroplasts (double arrowheads) are evident near the edge of the petiole.

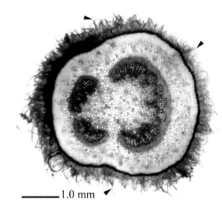

Figure 293. Wooly morning glory (*Argyreia nervosa*) petiole. In addition to the two arcs of vascular tissues, note the numerous covering trichomes (arrowheads).

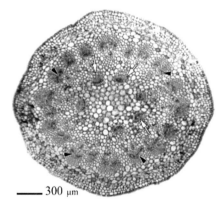

Figure 294. Umbrella tree (*Schefflera actinophylla*) petiole. Vascular tissues are arranged in an outer (arrowheads) and inner (arrows) ring.

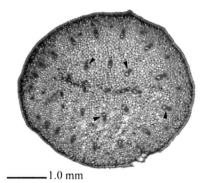

Figure 295. Sail plant; peace lily (*Spathiphyllum wallisii*) petiole. This monocot species has scattered vascular bundles (arrowheads) in its petiole.

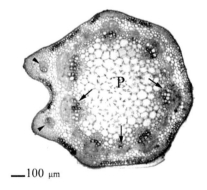

Figure 296. Grape (*Vitis vinifera*). The petiole contains a ring of vascular bundles (arrows) and two small vascular bundles (arrowheads). A broad pith (P) is present.

Chapter 10
Reproductive Organs

Flowers

There is enormous variation in the morphology of flowers and in the structure of their component organs. Often considerable structural information can be obtained by using rather thick sections viewed with a stereo binocular microscope. However, for detailed information at the cellular level, thinner sections of the various organs need to be examined using a compound microscope. In order to section petals and sepals, and sometimes stamens, it is often necessary to use a support such as Styrofoam (see **Chapter 2** for method). Frequently, considerable detail can be obtained from unstained sections mounted in water under a cover glass; in order to enhance contrast, however, sections can be stained with Toluidine blue O or other stains.

Flowers of three species are included to show the information that can be obtained using stereo binocular and compound microscopes.

Easter lily (*Lilium longiflorum*)

The large flowers of lily species, including Easter lily (*Lilium longiflorum*), are usually available for illustrating anatomical features of floral organs. The first and second whorls of organs are simliar in appearance and are therefore referred to as tepals. In Easter lily they can be referred to as petal-like tepals (Fig. 297).

Unstained thick sections of the tri-lobed **stigma** and a portion of the **style** retain numerous **pollen grains** on the stigma surface (Fig. 298). Sections of the stigma stained with Toluidine Blue O reveal the small papillae that develop as epidermal outgrowths (Fig. 299).

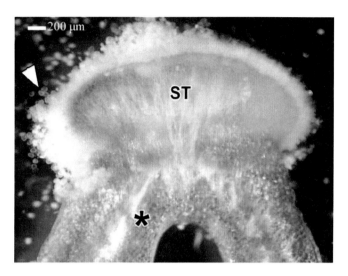

Figure 298. Thick longitudinal section of a stigma (ST) and part of a style (∗) of Easter lily (*Lilium longiflorum*) flower viewed with a stereobinocular microscope. Pollen grains (arrowhead) adhere to the stigma.

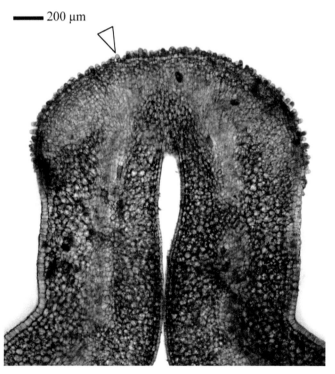

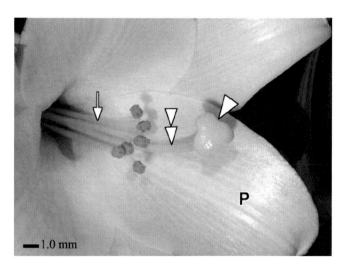

Figure 297. Flower of Easter lily (*Lilium longiflorum*) showing stigma (arrowhead), style (double arrowhead), stamens with anthers subtended by filaments (arrow), and petal-like tepals (P).

Figure 299. Longitudinal section of a stigma and style of Easter lily (*Lilium longiflorum*) flower stained with TBO. Numerous papillae (arrowhead) are present on the stigma surface.

Cross sections of the style reveal that it is hollow with **transmitting tissue** (tissue along which pollen tubes grow) lining the canal (Fig. 300). Three vascular bundles are present (Fig. 300). Sections of the **ovary** at various stages in development show the **locules**, developing **ovules**, and vascular tissue (Figs. 301, 302). The stigma, style, and ovary constitute the female part of the flower and are collectively referred to as a **gynoecium**.

Unstained transverse sections of the bilobed **anthers** show that each lobe consists of two **microsporangia** or **pollen sacs** (Fig. 303). At higher magnification, the layers of the anther and pollen grains are visible (Fig. 304). The reticulate nature of the **exine**, the outer layer of each pollen grain, is evident in samples stained with Toluidine Blue O (Fig. 305). A transverse section of the **filament** shows that a small vascular bundle is surrounded by parenchyma cells (Fig. 306). The anthers and filaments constitute the male part of the flower and are collectively referred to as **stamens**.

Petal-like tepals have stomata, loosely arranged **ground parenchyma,** and small veins (Figs. 307, 308).

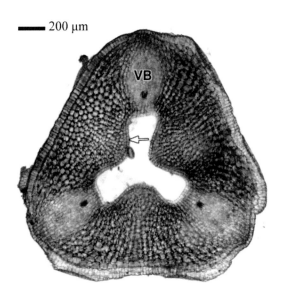

Figure 300. Transverse section of a style of Easter lily (*Lilium longiflorum*) flower stained with TBO. A central canal is surrounded by parenchyma cells (transmitting tissue—arrow). Three vascular bundles (VB) are present.

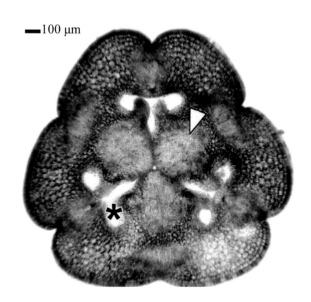

Figure 301. Transverse section of an ovary of Easter lily (*Lilium longiflorum*) flower stained with TBO showing locules (∗) with developing ovules (arrowhead).

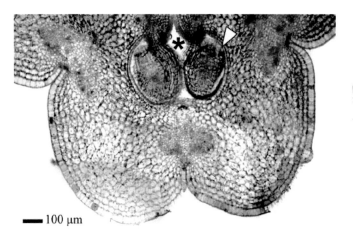

Figure 302. Transverse section of an ovary of Easter lily (*Lilium longiflorum*) flower stained with TBO showing a locule (∗) with developing ovules (arrowhead).

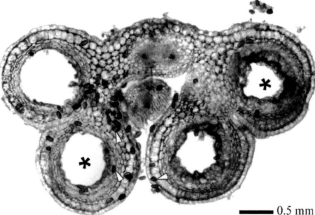

Figure 303. Unstained transverse section of an anther from an Easter lily (*Lilium longiflorum*) flower showing the two lobes, each containing two microsporangia or pollen sacs (∗). Pollen grains (arrowheads) have been dislodged by sectioning and lie over the surface of the anther.

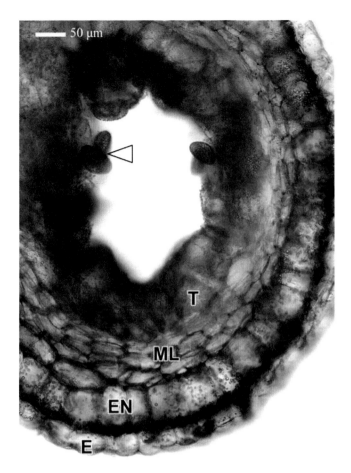

Figure 304. Transverse section of one pollen sac from an anther of an Easter lily (*Lilium longiflorum*) flower showing epidermis (E), **endothecium** (EN), middle layers (ML), and the **tapetum** (T). A few pollen grains are visible (arrowhead).

Figure 305. Pollen grains of Easter lily (*Lilium longiflorum*) stained with TBO showing the sculptured exine.

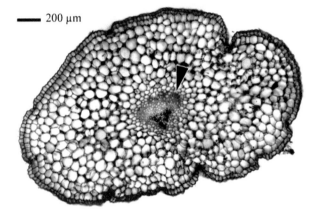

Figure 306. Transverse section of a filament of an Easter lily (*Lilium longiflorum*) flower stained with TBO. A single vascular bundle (arrowhead) is surrounded by parenchyma cells.

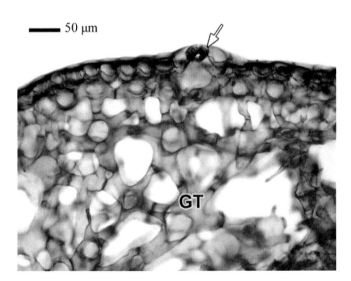

Figure 307. Transverse section of a petal-like tepal of an Easter lily (*Lilium longiflorum*) flower stained with TBO showing a stomatal complex (arrow) and loosely arranged ground tissue cells (GT).

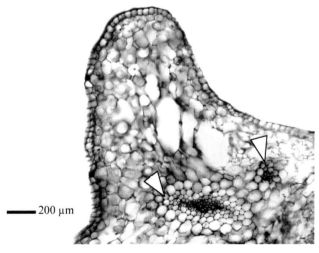

Figure 308. Transverse section of a petal-like tepal of an Easter lily (*Lilium longiflorum*) flower stained with TBO showing two small vascular bundles (arrowheads).

Peruvian lily (*Alstroemeria* sp.)

Alstroemeria, a genus of many species and cultivars, is usually available from florists. Like Easter lily, it is a monocotyledonous species in which the first two whorls of floral organs (each whorl consisting of three individual parts) are petal-like tepals (Fig. 309). The inflorescence is subtended by leaf-like **bracts** (Fig. 309). Each flower is bisexual; an elongated style terminated by a stigma consisting of three lobes (Fig. 310) and six stamens are visible with the naked eye.

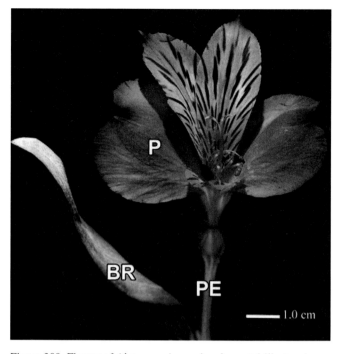

Figure 309. Flower of *Alstroemeria* sp. showing petal-like tepals (P), a bract (BR) that subtends the inflorescence, and the **pedicel** (PE).

Figure 310. Three stigmatic lobes (arrowhead) of an *Alstroemeria* sp. flower.

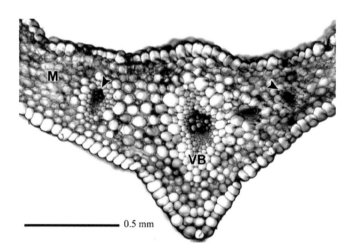

Figure 311. Transverse section of an *Alstroemeria* sp. bract stained with TBO. Unspecialized mesophyll (M), the midvein vascular bundle (VB), and smaller vascular bundles (arrowheads) are present.

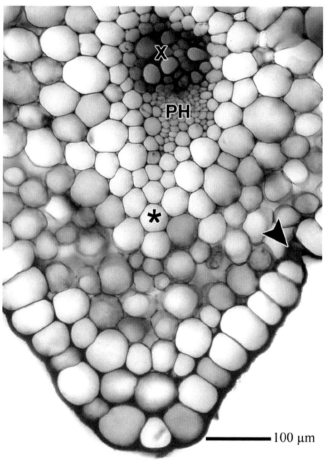

Figure 312. Higher magnification of the midvein of an *Alstroemeria* sp. bract stained with TBO. Xylem (X) and phloem (PH) are surrounded by midrib parenchyma (∗). A stomate (arrowhead) is evident.

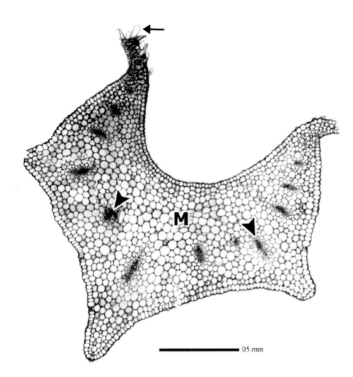

Figure 313. Transverse section of an *Alstroemeria* sp. petal-like tepal stained with TBO. Undifferentiated mesophyll (M), small vascular bundles (arrowheads), and unicellular trichomes (arrow) are evident.

Figure 314. Unstained top view of an *Alstroemeria* sp. anther with numerous small pollen grains (arrowheads).

Figure 315. Pollen grain of an *Alstroemeria* sp.

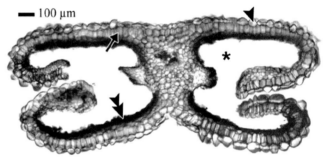

Figure 316. Transverse section of an *Alstroemeria* sp. anther stained with TBO. Pollen sacs (∗) that have lost their pollen grains, epidermis (arrowhead), endothecium (arrow), and remnants of the tapetum (double arrowhead) are evident.

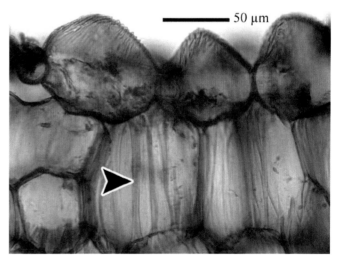

Figure 317. Endothecium cells of an *Alstroemeria* sp. anther stained with TBO showing lignified wall thickenings (arrowhead).

Sections of the bracts show the leaf-like anatomy of this organ in that photosynthetic undifferentiated mesophyll, a midrib with a single vascular bundle, and smaller vascular bundles are present (Fig. 311). Each vascular bundle consists of phloem and xylem that is demonstrated best in the midvein (Fig. 312). This vascular bundle is surrounded by considerable parenchyma composing the midrib (Fig. 312). Stomata are present in the epidermis (Fig. 312).

Thin sections of the petal-like tepals show the simple structure of these organs in that undifferentiated mesophyll and small vascular bundles are present (Fig. 313). Small, unicellular trichomes occur at the tips of each tepal adjacent to the stem (Fig. 313).

Mature anthers have numerous pollen grains (Fig. 314), the details of which can be seen when mounted in water under a cover glass and viewed with a compound microscope (Fig. 315). The four pollen sacs of mature anthers as well as the epidermis, endothecium, and remnants of the tapetum are evident in thin sections (Fig. 316). The endothecium of this species consists of cells with bands of lignified thickenings as shown by the blue staining with TBO (Fig. 317).

The simple anatomy of filaments is revealed in thin transverse sections; a central vascular bundle is surrounded by parenchyma cells and an epidermis (Fig. 318).

Thick transverse sections of the flower at the base of the stamens and style viewed with a stereo binocular microscope show sections of the filaments of the six stamens and the hollow style (Fig. 319). Thick longitudinal sections through the ovary show the numerous ovules within the locule (Fig. 320). Thin sections of the style viewed with a compound microscope show the transmitting tissue surrounding the stylar canal and three vascular bundles within parenchyma tissue (Fig. 321).

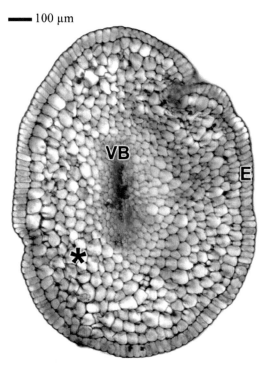

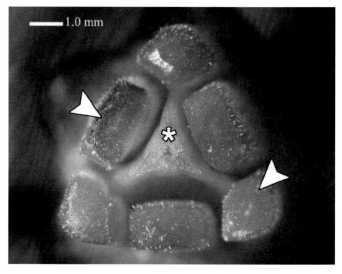

Figure 319. Thick transverse section at the level of the base of the filaments of an *Alstroemeria* sp. flower. Sections of six filaments (arrowheads) and the hollow style (below ∗) are evident.

Figure 318. Transverse section of an *Alstroemeria* sp. filament. A single vascular bundle (VB) is surrounded by parenchyma cells (∗) and an epidermis (E).

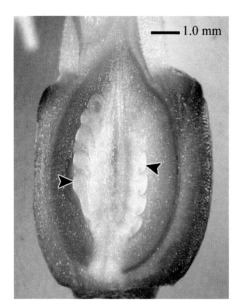

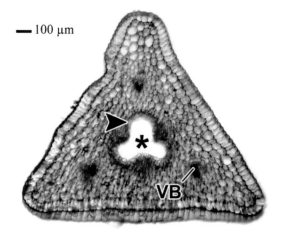

Figure 320. Thick longitudinal section through the ovary of an *Alstroemeria* sp. flower showing the numerous ovules (arrowheads) within a locule.

Figure 321. Transverse section of an *Alstroemeria* sp. style stained with TBO. The stylar canal (∗) is surrounded by transmitting tissue (arrowhead). Three vascular bundles (VB) are present.

Rose of China (*Hibiscus rosa-sinensis*)

There are numerous varieties of this species, many of which are grown as ornamentals and house plants. Most have basically the same floral structure with the exception of some cultivars that may have modified petals. One of the features of all *Hibiscus* species is the production of mucilage in most vegetative and floral parts that makes staining of sections somewhat difficult. Considerable information on floral structure can be obtained, however, with the use of rather thick sections viewed with a stereo binocular microscope.

Flowers of this dicotyledonous species consist of a series of bracts, five sepals, five petals, a gynoecium of five fused carpels, and a style terminating in a stigma consisting of five lobes, a column of fused stamen filaments (**staminal column**) surrounding the style, and many free anthers. Figure 322 shows the petals, stigmas, and some of the anthers of a flower. The five lobes of the stigma are shown in Figure 323 and an anther with pollen grains in Figure 324. The pollen grains are circular and have a highly modified, spiked exine (Fig. 325).

Figure 322. Flower of *Hibiscus rosa-sinensis* showing petals (P), the stigma lobes (arrowhead), and anthers (arrow).

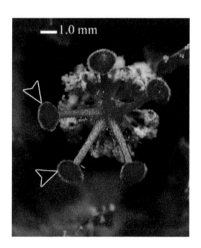

Figure 323. Stigma lobes (arrowheads) of *Hibiscus rosa-sinensis* flower.

Figure 324. Open anther of *Hibiscus rosa-sinensis* flower with pollen grains (arrowhead).

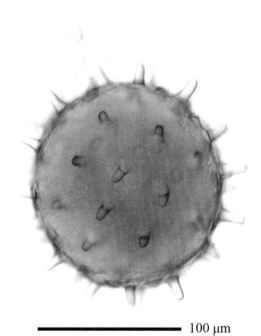

Figure 325. Pollen grain of *Hibiscus rosa-sinensis*.

Longitudinal sections of the flower show the whorls of floral organs: bracts, sepals, petals, the staminal column, the style and ovary of the gynoecium (Fig. 326). An enlargement of the ovary shows the numerous ovules present (Fig. 327).

Transverse sections at the base of the sepals and petals show the arrangement of these organs surrounding the gynoecium (Fig. 328). Transverse sections of the ovary show the five locules each with two ovules (Fig. 329).

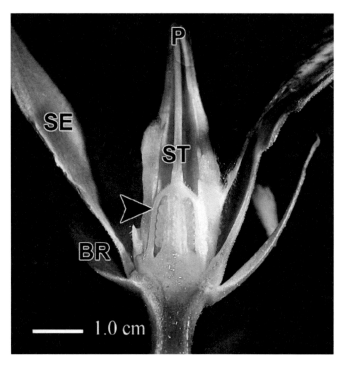

Figure 326. Thick longitudinal section of *Hibiscus rosa-sinensis* flower showing sections of bracts (BR), sepals (SE), petals (P), and the style (ST) and ovary (arrowhead) of the gynoecium.

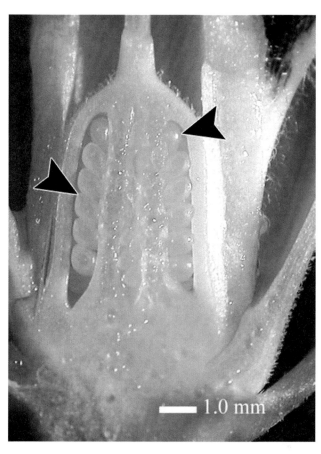

Figure 327. Thick longitudinal section of *Hibiscus rosa-sinensis* flower showing the ovary with many ovules (arrowheads).

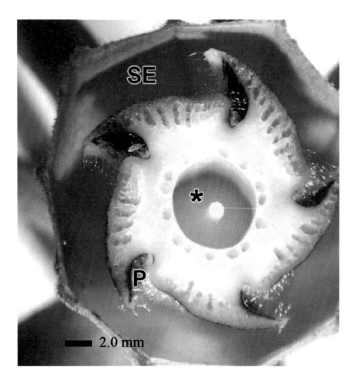

Figure 328. Thick transverse section at the base of the petals (P) and sepals (SE) of an *Hibiscus rosa-sinensis* flower showing the arrangement of these organs surrounding the gynoecium (∗).

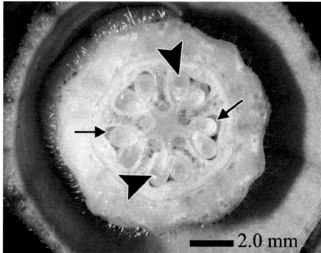

Figure 329. Thick transverse section of the ovary of an *Hibiscus rosa-sinensis* flower showing the five locules (arrows) with ovules (arrowheads).

Thin transverse sections of the bract stained with TBO and mounted directly in water show the copious amount of mucilage produced by cells in this organ (Fig. 330). Removing the mucilage by rubbing the surface of sections with a fine brush before mounting in water allows the structure of the bract to be determined. The green bracts have photosynthetic, undifferentiated mesophyll cells, small vascular bundles, and a single layer of epidermal cells (Fig. 331). Xylem and phloem are easily seen at higher magnification of the central vascular bundle (Fig. 332).

Figure 330. Thin transverse section of a bract of an *Hibiscus rosa-sinensis* flower stained with TBO surrounded by copious mucilage (∗).

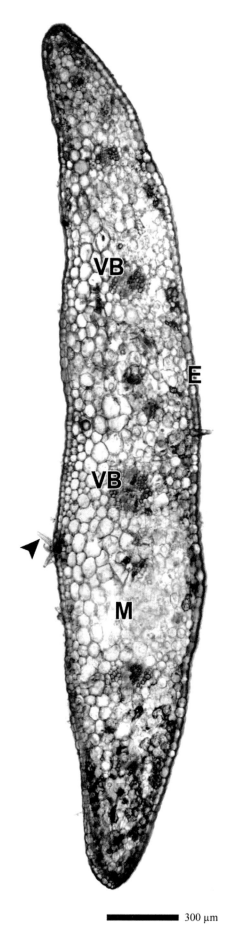

Figure 331. Thin transverse section of a bract of an *Hibiscus rosa-sinensis* flower stained with TBO after removal of the mucilage. Photosynthetic mesophyll (M), small vascular bundles (VB), and the epidermis (E) with a trichome (arrowhead) are evident.

Thin unstained sections of sepals show the simple anatomy of these organs; photosynthetic undifferentiated mesophyll, small vascular bundles, and a single layer of epidermis with numerous branched covering trichomes are present (Fig. 333). The structure of individual trichomes on sepals is illustrated in Figure 334.

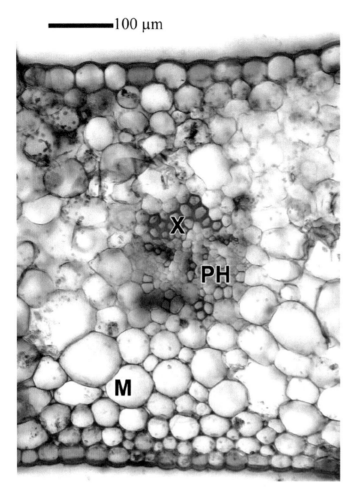

Figure 332. Mesophyll (M) and a vascular bundle with xylem (X) and phloem (PH) in a thin transverse section of a bract of an *Hibiscus rosa-sinensis* flower stained with TBO after removal of the mucilage.

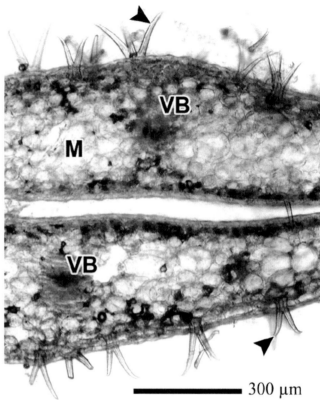

Figure 333. Unstained thin section of sepals of an *Hibiscus rosa-sinensis* flower showing photosynthetic mesophyll (M), small vascular bundles (VB), and an epidermis with numerous branched covering trichomes (arrowheads).

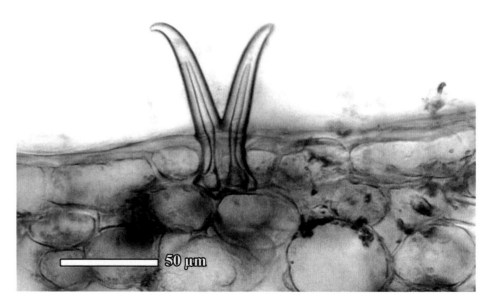

Figure 334. Branched covering trichome on a sepal of an *Hibiscus rosa-sinensis* flower.

Box 24. Demonstration of germinating pollen grains

During the reproductive cycle of flowering plants, pollen grains produced in anthers germinate once they are transferred to an appropriate stigma of a flower either by wind or by an insect or other animal vector. During this process, water is imbibed by the pollen grain, triggering the outgrowth of one or more **pollen tubes** that grow down the style to reach ovules located in the ovary. Each pollen tube carries two male gametes (sperms), one of which will fertilize the egg within an ovule. Pollen grains of a few plant species can be germinated artificially by providing them with water; most, however, require a nutrient solution. Pollen grains of *Impatiens* spp. show rapid germination once they are imbibed in water.

A simple method to demonstrate this is to begin by removing pollen grains from mature anthers. Place pollen grains in a drop of water on a glass slide and then apply a cover glass. By viewing the slide at low magnification with a compound microscope, pollen grains showing the initial protrusion of a pollen tube can be located. At higher magnification, these should be observed over a period of a few minutes to study the elongation of the pollen tube and cytoplasmic streaming within the tube.

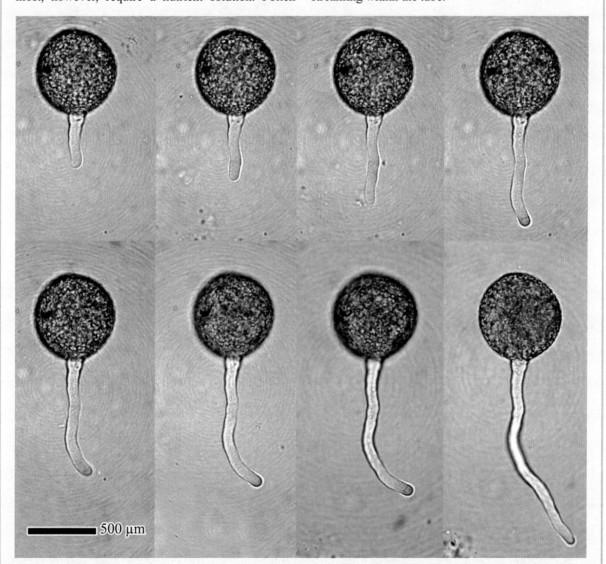

Time sequence (1–5 minutes) of pollen tube growth from germinated *Impatiens* sp. pollen grains after immersion in water.

Embryos

Arabidopsis thaliana

Developing siliques (fruits) of this species can be used to study embryogenesis by using the method outlined in **Box 25**.

Several stages, including the globular stage (Figs. 335, 336), the early and later stages in cotyledon formation (Figs. 337, 338), and the mature embryo stage (Fig. 339), are illustrated from ovules cleared and viewed with Nomarski or DIC optics.

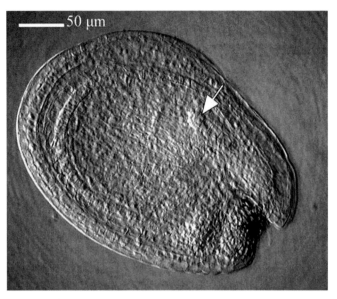

Figure 335. Cleared ovule of *Arabidopsis thaliana* showing globular stage (arrow) of embryo development.
Photo courtesy of Ryan Geil.

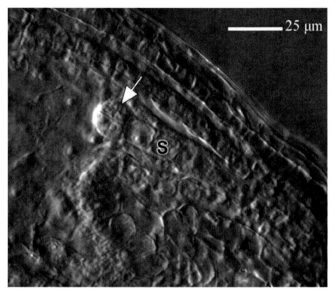

Figure 336. Cleared ovule of *Arabidopsis thaliana*; the elongated suspensor (S) subtends the developing embryo (arrow).
Photo courtesy of Ryan Geil.

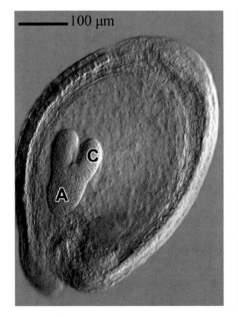

Figure 337. Cleared ovule of *Arabidopsis thaliana* showing the formation of the cotyledons (C) and the embryo axis (A) in a developing embryo. Photo courtesy of Ryan Geil.

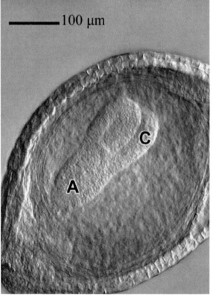

Figure 338. Cleared ovule of *Arabidopsis thaliana* showing a later stage in the formation of the cotyledons (C) and the embryo axis (A) in a developing embryo.
Photo courtesy of Ryan Geil.

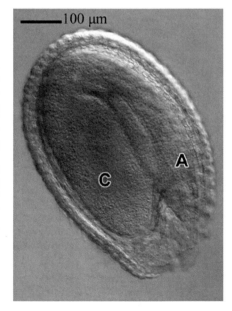

Figure 339. Mature ovule of *Arabidopsis thaliana*. The cotyledons (C) and embryo axis (A) occupy most of the volume of the developing seed. Photo courtesy of Ryan Geil.

Box 25. Clearing *Arabidopsis* ovules with Hoyer's solution

This protocol can be used for clearing intact seeds of *Arabidopsis thaliana*, which can then be observed using either Nomarski or Differential Interference Contrast — DIC optics. This is an efficient way to analyse embryo and endosperm development without sectioning.

1. To prepare a modified Hoyer's solution, mix:
    a. 7.5 g gum arabic (Sigma, G9752)
    b. 100 g chloral hydrate (Sigma, C8383; controlled substance)
    c. 5 mL glycerol (Sigma, G6279)
    d. 60 mL water
    e. Dissolve by stirring 3–5 hours, or overnight.
2. Open the siliques (seed pods) with a pair of forceps and a sharp needle; pick up the intact ovules with the needle.
3. Put a drop of Hoyer's solution on a slide.
4. For each sample, place 20–30 ovules into the Hoyer's solution.
5. Gently add a cover glass to the sample.
6. Keep the slides at room temperature for times ranging from 15 minutes to overnight, depending on the developmental stages of the ovules.
7. Observe the samples with a microscope equipped with Nomarski or DIC optics.

Reference: Liu, C.M., and Meinke, D.W. 1998. The *titan* mutants of *Arabidopsis* are disrupted in mitosis and cell cycle control during seed development. Plant Journal 16: 21–31.

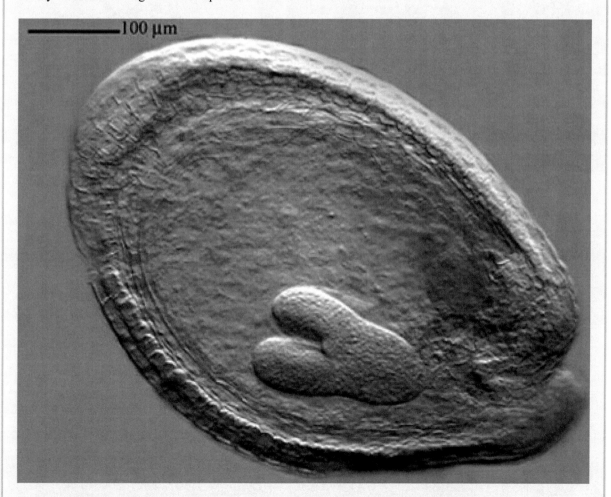

Cleared *Arabidopsis thaliana* ovule showing a developing embryo. Photo courtesy of Ryan Geil.

APPENDICES

Appendix 1. List of plant anatomy texts

Beck, C.B. 2005. An Introduction to Plant Structure and Development. Cambridge University Press, Cambridge.

Cuttler, D.F., Botha, T., and Stevenson, D.W. 2007. Plant Anatomy. An Applied Approach. Blackwell Publishing, Malden, Mass.

Cutter, E.G. 1978. Plant Anatomy. Part 1. Cells and Tissues. 2nd Ed. Edward Arnold (Publishers) Ltd., London.

Cutter, E.G. 1971. Plant Anatomy. Part 2. Organs. Edward Arnold (Publishers) Ltd., London.

Dickison, W.C. 2000. Integrative Plant Anatomy. Harcourt Academic Press, San Diego.

Esau, K. 1953. Plant Anatomy. John Wiley & Sons, Inc., New York.

Esau, K. 1977. Anatomy of Seed Plants. 2nd Ed. John Wiley & Sons Inc., New York.

Evert, R.F. 2006. Esau's Plant Anatomy. Meristems, Cells, and Tissues of the Plant Body—Their Structure, Function, and Development, 3rd Ed. John Wiley & Sons, Inc., Hoboken, N.J.

Fahn, A. 1990. Plant Anatomy. 4th Ed. Pergamon Press, Oxford.

Mauseth, J.D. 1988. Plant Anatomy. The Benjamin/Cummings Publishing Company, Inc., Menlo Park, Calif.

O'Brien, T.P., and McCully, M.E. 1969. Plant Structure and Development. A Pictorial and Physiological Approach. The Macmillan Company, Toronto.

Appendix 2. Preparation and use of stains

Reasons for staining plant material

Although organelles such as chloroplasts and chromoplasts are pigmented and therefore can be seen without staining, most organelles and other plant structures are colourless and can be seen more easily when stained. There are many general stains that can be used to highlight various components of plant cells.

Some stains are used as histochemical tests in which the colour reaction produced indicates the chemical nature of the substance stained. For example, if a small, roundish, colourless body is observed in an unstained section, various histochemical tests can be used to determine its nature. I_2KI can be used to determine whether it is a starch grain and a number of stains could be used to determine whether it is a protein body or a lipid body. When performing a histochemical test, it is important to also observe unstained material as a control to confirm the colour reaction of the stain. Control sections should be treated in the same manner as stained sections with respect to solvents, pH, etc., that are part of the staining procedure.

*In the list of stains below, those marked with an asterisk can be used as histochemical tests because of their specificity for particular chemical substances in plant cells.

Caution: Treat all stains as though they may be harmful. Avoid contact with skin, and breathing in dust or vapour from solutions.

Acetocarmine: Add 45 mL glacial acetic acid to 55 mL distilled water and then add 0.5 g carmine. Boil gently for 5 minutes, shake, cool, and filter through Whatman filter paper. Tissue should be placed in the stain and warmed gently. Nuclei should stain a deep red colour.

*Acid fuchsin: Dissolve 1.0 g acid fucshin in 100 mL water. Tissue can be mounted directly in the stain. Protein should stain pink–red.

*Aniline blue: Dissolve 5.0 mg water-soluble aniline blue in 100 mL K_2HPO_4 at pH 8.2. The solution will change from blue to colourless after a few hours. Sections can be mounted directly in the stain and viewed by fluorescence microscopy. Callose will appear yellow when viewed with a filter set providing blue light, and a silvery-blue when viewed with a UV filter set.

Reference: Currier, H.B. 1957. Callose substance in plant cells. American Journal of Botany 44: 478–488.

*Aniline blue black: Dissolve 1.0 g aniline blue black in 100 mL 7% acetic acid. Tissue should be stained for 5–10 minutes and then rinsed with 7% acetic acid. Protein bodies should stain dark blue.

Berberine hemisulphate: Dissolve 0.1 g berberine hemisulphate in 100 mL water. Hand-sectioned material should be stained for 15–30 minutes, rinsed several times with water, and mounted in a solution of 0.1% ferric chloride in 50% glycerol. Observe with either blue or UV light. Suberin, lignin, and other phenolic-rich substances will fluoresce. For improved definition of minute structures such as Casparian bands in roots, follow the berberine staining and rinsing with a counterstain of 0.5% aniline blue (water-soluble) for 30 minutes or 0.05% Toluidine Blue O for 0.5–3 minutes.

Reference: Brundrett, M., Enstone, D., and Peterson, C.A. 1988. A berberine – aniline blue fluorescent staining procedure for suberin, lignin, and callose in plant tissue. Protoplasma 146: 133–142.

*Cellufluor (formerly called Calcofluor White M2R): Prepare an aqueous solution (10 mg in 100 mL). Heat gently to dissolve stain. Hand-sectioned material needs to be stained for only 1–2 minutes, then washed with water before mounting in water. Viewed with UV light, cellulosic walls should fluoresce a blue–white.

Chlorazol Black E: Dissolve 30 mg Chlorazol Black E in 33 mL water, and then add 33 mL of lactic acid and 33 mL of glycerol. The solution may need to be filtered through Whatman filter paper before use. This stain is useful for visualizing veins in cleared leaves and Casparian bands in the endodermis and exodermis of cleared roots. Stained structures should appear black.

*DAPI (4'-6-diamidino-2-phenylindole): Prepare the staining solution at a concentration of 0.1 μg/mL in PBS (Phosphate Buffered Saline) buffer and adjust the pH to > 7. Whole mounts of thin leaves or hand-sections can be used to demonstrate nuclei. If whole mounts of leaves are used they should be incubated in 50% ethanol for 10 minutes so that the stain will penetrate into the cells. Tissue should be washed with water after this treatment. Mount the material in the stain and let stand about 5 minutes. Specimens must be viewed with UV light. Nuclei will fluoresce a silvery–blue colour.

PBS buffer: Prepare the following stock: 4 g potassium chloride, 160 g sodium chloride, 4 g monobasic potassium phosphate (KH_2PO_4), 43.2 g dibasic sodium phosphate ($Na_2HPO_4 \cdot 7H_2O$), distilled water to 1 L. This must be diluted 20× before use.

*Iodine potassium iodide (I_2KI): Dissolve 2 g potassium iodide (KI) in 100 mL water, and then dissolve 0.2 g iodine in the KI solution. Starch grains should appear blue to almost black within a few minutes of introducing the stain.

Caution: Handle iodine with care since it is poisonous. Do not breathe iodine vapours.

*Phloroglucinol-HCl: Prepare a saturated solution of phloroglucinol-HCl in 20% hydrochloric acid (HCl) and filter before use. Tissue must be mounted directly in the stain for viewing. Lignified cell walls will appear various shades of red, the intensity depending on the type and amount of lignin.

Caution: Do not get the stain or fumes on the objectives of the microscope since HCl will etch the lenses. Make sure that the stain does not leak out from under the cover glass.

Rhodamine 123 as a stain for mitochondria: This is a fluorescent probe that is also a vital stain, i.e., can be used to test whether cells are dead or alive. It is useful to demonstrate the occurrence of mitochondria in plant cells. Prepare a stock solution by adding 1.0 mg stain to 1.0 mL distilled water. Dilute stock to 10 μg/mL in water. Immerse tissue in dye for 0.5–2 hours. Isolated cells may be stained for 10–20 minutes. Rinse in distilled water and mount in water under a cover glass. Observe under blue light. Minimize time of exposure to the excitation light, as the dye is prone to photobleaching.

Mitochondria should fluoresce green under blue excitation. If chloroplasts are present, chlorophyll should fluoresce red.
Reference: Wu, F.-S. 1987. Localization of mitochondria in plant cells by vital staining with rhodamine 123. Planta 171: 346–357.

*Sudan stains: There are a number of formulations of Sudan stains that can be prepared. A saturated solution of either Sudan III or Sudan IV (we often use a combination of both) in 70% ethanol should be filtered before use. Sudan red 7B (fat red) can also be used and gives excellent results for suberin and cutin. Stain tissue for 5 or more minutes, adding more stain as necessary. Rinse with 50% ethanol. Sections are best mounted in 50% glycerol for viewing. Lipids (including oils that occur in the cytoplasm) will stain orange to red. The cuticle, cutinized cell walls, and walls containing suberin will also stain orange to red.

Toluidine Blue O: This is a very useful general stain for studying plant tissues, since it is water-soluble and produces a range of colours depending on the binding sites in the tissue and the pH of the dye solution. Such a dye is referred to as **metachromatic**. Cell walls containing pectic substances stain pink to reddish purple, while those containing lignin stain various shades of blue or blue green. Nuclei stain blue to greenish blue, and various phenolic compounds may also stain shades of blue.

Dissolve 50 mg Toluidine Blue O in 100 mL water (or in citrate buffer, pH 4.0 for the best metachromicity). The protocol for staining is given in **Box 5**.

Citrate buffer: Prepare stocks of 0.1 M citric acid (A) and 0.1 M sodium citrate (B). For citrate buffer of pH 4.0 use 25.5 mL (A) and 24.5 mL (B) and top up to 100 mL with distilled water. Adjust the pH to 4.0 by adding drops of 1N HCl.

*Vanillin-HCl: This is used as a simple staining reaction for some tannins and polyphenolic substances. Add 2.5 g vanillin (4-hydroxy-3-methoxybenzaldehyde) to 25 mL ethanol. Prepare the final staining solution by adding 10 mL concentrated HCl to 10 mL of the vanillin stock solution. **Care must be taken when handling the HCl solution; always add the acid to the stock solution.** Stain fresh sections for 10–60 seconds and then mount them under a cover glass. Many phenolics will stain red or pink.

Basic reference books for staining plant tissues

Berlyn, G.P., and Miksche, J.P. 1976. Botanical Microtechnique and Cytochemistry. Iowa State University Press, Ames, Iowa.

Gahan, P.B. 1984. Plant Histochemistry and Cytochemistry. An Introduction. Academic Press, London.

Jensen, W.A. 1962. Botanical Histochemistry. W.H. Freeman and Company, San Francisco.

O'Brien, T.P., and McCully, M.E. 1981. The Study of Plant Structure. Principles and Selected Methods. Termarcarphi Pty. Ltd., Melbourne.

Ruzin, S.E. 1999. Plant Microtechnique and Microscopy. Oxford University Press, Inc., New York.

Appendix 3. Some unwelcome intruders and problems when sectioning

Dust

The bane of microscopists, dust particles are often seen with the microscope. These particles are opaque bodies with irregular edges and, of course, come in all sizes.

In research laboratories, microscopes are usually housed in rooms that are kept as free from dust as possible. Unfortunately, this is not practical for high school and university undergraduate student laboratories.

If dust particles are infrequent and not overlying an object of interest, and no pictures are to be taken, the particles can just be ignored. On the other hand, the presence of copious particles may mean that the microscope has not been properly cleaned. Before dust can be removed, it must first be located, and this takes some detective work, as it may be present on one or more of several surfaces.

Is the dust somewhere on or in the eyepieces?

Dust in these locations is easy to detect by rotating each eyepiece in turn while looking through it. If dust particles are located there, they will rotate along with the eyepiece. There is often dust on the upper surface of the eyepiece. Clean this surface and try the rotation test again. If dust is still there, slide the eyepiece out of its tube, cover the open end of the tube to prevent dust from entering the microscope, and clean the innermost lens. If the dust is still there, the eyepiece will need to be taken apart and cleaned by a professional. However, most often dust is located on the outer surface of the lens.

Is the dust on an objective lens?

This can be checked by rotating another objective lens into place. If the dust disappears, then it was located on the original lens. Clean the outer surface of this lens. Rarely, dust may be on the internal surface of the lens. It can be screwed out of the microscope and the uppermost face cleaned. As before, if cleaning these two surfaces is ineffective, then the lens will need to be dismantled and cleaned professionally.

Is the dust on the microscope slide?

Move the slide around while looking at it through the eyepieces. If the dust also moves, it is on the slide. Depending on the mounting material, it may be possible to pry off the cover glass and replace it with a new one. As a preventative measure, keep boxes of slides and cover glasses closed when not in use. Dusty slides or cover glasses can be cleaned by dipping them in alcohol before use.

Is the dust on the condenser?

Using the condenser adjustment knob, raise and lower the condenser while looking through the eyepieces. If the dust comes in and out of focus, then it is located on the condenser, usually on the top surface, which can easily be cleaned. If this does not remove the dust, try cleaning any other accessible part of the condenser that is in the light path. If it is not possible to get rid of the dust, a temporary solution to the problem is to lower the condenser so that the dust is not in focus while the microscope is being used.

Is the dust on the iris diaphragm?

Dust in this location can be seen with the naked eye when the microscope light is on and is easily removed.

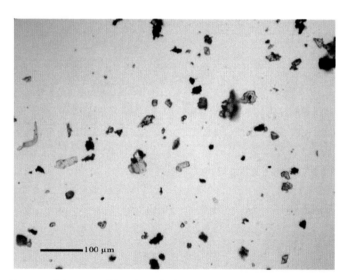

Dust particles on a slide, viewed with a compound microscope.

Lint

After dust, lint is probably the most common "foreign" material seen with microscopes. This material originates from fragments of cloth, bench-lining, absorbent paper, and cotton swabs — sometimes the very material used to clean microscope lenses! Like dust, lint is small and light enough to be suspended in the air and move around by air currents. It is also opaque but its shape is long and narrow. To locate and remove lint, follow the procedures for dust given above.

Air

Bubbles of air often form when a specimen is being mounted in liquid on a slide. They are characterized by a thick, dark border and are often the most conspicuous objects to be seen, leading some students to identify them as cells. When they occur away from the specimen, they will appear as regular circles but when they form over, under, or around the specimen, their shape can be modified by contact with it.

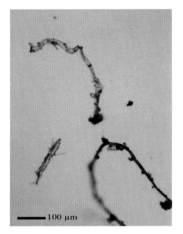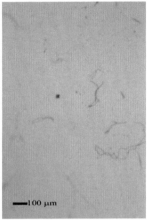

Lint on microscope slide (left) and on condenser (right).

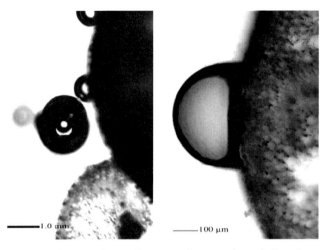

Air bubbles in mounting medium away from specimen (left) and adhering to the specimen (right).

Chips of glass

This is not a common problem but sometimes the edges of cover glasses or slides will chip and small fragments become mixed with the material being viewed. Such chips are rather transparent but have irregular shapes with sharp-angled edges. If this is a problem, rinse the slides and cover glasses with alcohol before use.

The presence of air on a slide is not always an artefact of preparation. Most plant tissues have air spaces between their cells and, under normal conditions, these contain a gas mixture. Sectioning the tissue tends to flood the spaces, but does not always remove this gas.

Sometimes air bubbles can be removed from a wet mount by raising one side of the cover glass and moving it up and down rapidly. Creation of air bubbles can be minimized by

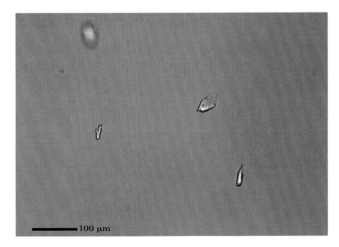

Glass chips on microscope slide.

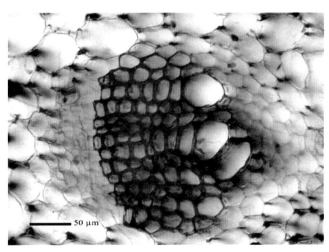

Air bubbles appear as black spots in intercellular air spaces.

always putting a drop of the liquid mounting material on the slide first, then slipping the sections into it, and finally lowering the cover glass slowly.

A feature of wet mounts is that they dry out, drawing air under the cover glass and, with time, around and over the specimen. A meniscus between the air and mountant is visible with the naked eye. More liquid can be added to the slide by placing a small drop so that it touches the edge of the cover glass. It will be drawn under the cover glass by capillary action. Formation of air bubbles during slide rehydration can be avoided if the drop is added so as to touch the edge of a meniscus.

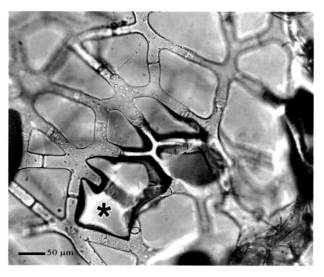

When aerenchyma tissue is sectioned, large air bubbles (∗) often form over and around the cells, and degrade the image.

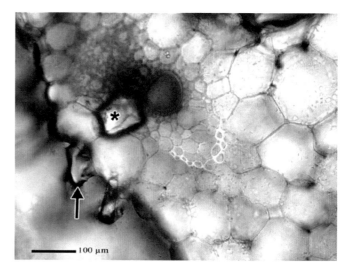

Air bubble (∗) overlying section. Liquid–air meniscus (arrow) at edge of section.

Spiral walls of tracheary elements

Students are often intrigued to see objects resembling springs or coils protruding from, or overlying, their sections. These are spirally thickened, lignified walls of vessels or tracheids that have been caught with the razor blade during sectioning and have been pulled across the section.

Oblique sections

A common problem encountered when preparing sections of fresh material is that sectioning has occurred at an oblique angle (i.e., not at 90° to the axis of the organ), causing the tissue to appear smeared. This makes the interpretation of cellular organization very difficult. This problem can be overcome by making sure that sectioning is done at right angles to the long axis of the specimen as illustrated in **Chapter 2**.

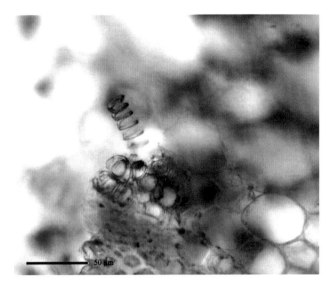

Spirally thickened tracheary element.

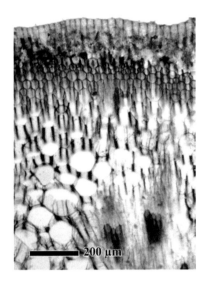

Oblique section of *Alstroemeria* sp. stem stained with TBO.

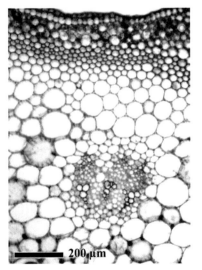

Non-oblique section of *Alstroemeria* sp. stem stained with TBO.

Appendix 4. A simple method for staining cleared roots and leaves

Example: Demonstrate the colonization of roots by arbuscular–mycorrhizal fungi

1. Clear roots in 5% KOH at 90 °C for 1 hour or more.

2. Rinse in water and stain in 5% ink (Schaeffer black ink gives the best results) in vinegar (or 5% acetic acid diluted in water) for a few minutes at 55 °C.

3. Rinse in distilled water and if roots are overstained, allow them to destain in water containing a few drops of vinegar (or 5% acetic acid diluted in water).

4. Mount in water or 50% glycerine on a microscope slide and cover with a cover glass.

Staining cleared leaves

Use the same protocol as above, but note that staining of cleared leaves with ink may take up to 2 hours depending on the type of leaf.

Reference

Vierheilig, H., Coughlan, A.P., Wyss, U., and Piché, Y. 1998. Ink and vinegar, a simple staining method for arbuscular–mycorrhizal fungi. Applied and Environmental Microbiology 64: 5004–5007.

Appendix 5. Testing viability of plant cells

There are many tests to determine whether individual cells are dead or alive (see Stadelmann and Kinzel 1972; Ruzin 1999). The three examples given below test whether the outer membrane of the cells (i.e., plasmalemma or plasma membrane) has retained its selectively permeable properties.

The inner epidermis from an onion bulb scale is a convenient material to illustrate the results of viability tests. To prepare many epidermal strips, remove an onion bulb scale and cut cross-hatchings into its inner face with a sharp razor blade. With forceps, gently peel a square of the epidermis off and mount it, cuticle-side-up, in the desired solution (**refer to Figures 15 and 16 in Chapter 3**). Best staining results are obtained with peels from relatively fresh bulbs. If the epidermis does not lift off easily, the scale can be vacuum infiltrated with water prior to peeling. Most cells in the middle of a peel should be alive, but all cut cells around the edge will be dead. To kill all the cells in a peel, place it in boiling water for 5 minutes. Let the peel rest for 1 hour to allow time for the plasma membrane to degrade before testing for viability.

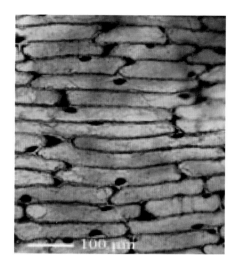

Epidermal peel of onion bulb scale. Cells in the peel were killed by heating prior to staining with Evans blue. Note blue colour in cytoplasms and nuclei.

Deplasmolysis test

This test is considered to be an excellent indicator of cell vitality. First a cell is plasmolyzed by placing it in a sufficiently strong solution of mannitol (usually a 0.3 M solution is adequate) for 30 minutes. Then the solution is changed back to water by drawing off the mannitol on one side of the cover glass and replacing it with water added to the other side. The cells should rest for a 30 minute period. Living cells will deplasmolyze (i.e., the protoplast will expand to fill the cell), whereas dead cells will not. This test works best on cells with a granular (and thus easily visible) cytoplasm.

Evans blue test

Evans blue is a non-permeating stain (Taylor and West 1980). A damaged plasma membrane (of a dead cell) will allow the stain to pass into the cytoplasm, and the nucleus and cytoplasm will stain blue. However, an intact membrane (of a living cell) will not allow the stain to pass, so that the cytoplasm and nucleus remain unstained. Note: this test is not suitable for tissues that have been dead for a long time during which the nucleus and cytoplasm may have been removed by micro-organisms.

Apply a staining solution (0.5% Evans blue in distilled water) to the specimen for 15 minutes. Rinse in water until no more stain leaves the section. Mount in water on a slide and observe cells for stained contents.

Uranin (disodium fluorescein) test

This test works by an ion trap mechanism. It requires plasma membrane and tonoplast integrity, and a cytoplasmic

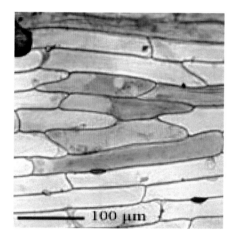

Epidermal peel of onion bulb scale. Both living and dead cells are present. Living cells have not stained with Evans blue. Dead cells have stained blue.

pH near 7. In solution, the sodium ions and the fluorescein dissociate. At low pH, most of the fluorescein molecules are protonated (i.e., with hydrogen ions replacing the sodium ions). In this form, the molecules are electrically neutral and will diffuse through the lipid bilayer of the plasma membrane. Once in the cytoplasm, a molecule encounters a higher pH, loses its hydrogen ions, becomes negatively charged, and becomes membrane impermeant. Thus, the stain accumulates in the cytoplasm. It is usually even more strongly accumulated in nuclei and these are the first parts of the protoplast in which the fluorochrome can be seen. Dead cells do not accumulate fluorescein and remain dark under blue light.

Soak the specimens in 0.01% uranin in 0.07 M KH_2PO_4 (or any dilute, non-toxic buffer of pH 5.3 for 10 to 30 minutes. Rinse three times with the buffer used to make up the stain, and mount in buffer on a slide. Observe with violet or blue light using a fluorescence microscope. Filter sets made specifically for fluorescein are available. With short staining times, uranin accumulation in the nuclei and cytoplasms of cells can be seen. However, with longer staining or rinsing times, dye is transferred to the vacuoles.

There are some disadvantages in using uranin. Some cells that are clearly living (showing cytoplasmic streaming, for example) do not stain with uranin for unknown reasons. Therefore, it is necessary to do some preliminary testing on the subject material to see if it will stain reliably with this method.

Other forms of fluorescein can be used to test cell vitality. In addition to the requirements for uranin described above, fluorescein diacetate (FDA) requires that the cytoplasm contain esterases. FDA has the advantage of being nonfluorescent until it reaches the cytoplasm where its acetates are removed, liberating the fluorochrome. Thus, it is not necessary to rinse the staining solution from the specimen before observing it. Additions to the fluorescein molecule (e.g., halides) make it less membrane-permeable after being trapped in the cytoplasm.

Neutral red test

1. Prepare a 0.01% solution of neutral red in 0.08 M phosphate buffer pH 7.2. Note: for this technique to work, the neutral red must be applied to the tissue in a non-toxic buffer at pH 7 or higher. The chemical is not highly water soluble at this pH, so it is necessary to prepare the buffer first and add the neutral red solid to it slowly. The chemical will tend to precipitate out of solution, so it needs to be made up just before use.
2. Pour the neutral red solution into a small dish and submerge the specimens to be tested in the liquid. After 30 min, rinse the specimens in the buffer and mount them in buffer on a slide.
3. The vacuoles of living cells will be coloured brick red due to the accumulation of neutral red; vacuoles of dead cells will remain clear. Note: this method does not work equally well with all cells. It is required that the cytoplasm be at a neutral pH and the vacuole at an acidic pH, and the plasma membrane and tonoplast must be intact.

References

Ruzin, S.E. 1999. Plant Microtechnique and Microscopy. Oxford University Press Inc., New York.

Stadelmann, E.J., and Kinzel, H. 1972. Vital staining of plants cells. *In* Methods in Cell Physiology. *Edited by* V.D.M Prescott. Academic Press, New York pp. 325–372.

Taylor, J.A., and West, D.W. 1980. The use of Evans blue stain to test the survival of plant cells after exposure to high salt and high osmotic pressure. Journal of Experimental Botany 3: 571–576.

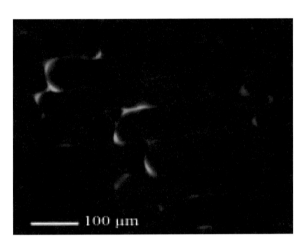

Epidermal peel of onion bulb scale stained with uranin for 10 minutes. Brightly fluorescent uranin is accumulated in the cytoplasm, which is typically thicker in the corners of cells.

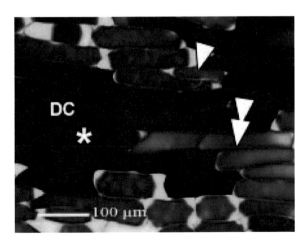

Epidermal peel of onion bulb scale stained with uranin for 1 hour. In some cells, uranin is still in the cytoplasm (∗). Transfer to the vacuole has begun (arrowhead), and in others, transfer to the vacuoles is complete (double arrowhead). Dead cells (DC) appear black.

Appendix 6. Suppliers

Type "science supplies" into an online search engine for a range of options.

Electron Microscopy Sciences

(www.emsdiasum.com)

A comprehensive selection of stains in powder form.

Sigma-Aldrich

(www.sigmaaldrich.com)

Produces and sells a broad range of biochemicals, organic and inorganic chemicals and related products.

Fisher Scientific

(www.fishersci.com)

Sells a complete range of chemicals and lab supplies.

Fisherbrand* Plain glass microscope slides.

Fisherbrand* Cover glasses.

Indigo Instruments

(www.indigo.com)

Source of microscope slides and cover glasses.

Home Science Tools

(http://www.hometrainingtools.com)

Basic slides and supplies.

Edmund Scientific

(http://scientificsonline.com)

Basic scientific supplies for home and school.

Basic Science Supplies

(www.basicsciencesupplies.com)

Electron Microscopy Sciences

(www.emsdiasum.com)

Source of a wide selection of razor blades.

Vanguard Steel

(www.vanguardsteel.com)

Catalogue No. 48070060

Persona super double edged razor blades.

GLOSSARY

accessory cells: specialized cells of the epidermis adjacent to guard cells; also called subsidiary cells. They differ in shape and function from other epidermal cells

adventitious root: a root that arises from an organ other than the embryo or other roots, usually stems

aerenchyma: a type of modified parenchyma tissue with large air lacunae (spaces)

aerial root: a root that grows above ground

amyloplast: a colourless plastid that stores starch

angular collenchyma: a type of collenchyma without intercellular air spaces; the primary cell wall is unevenly thickened

annular cell wall thickening: secondary cell wall deposited as a series of rings in tracheary elements

anther: that part of a stamen in which pollen grains develop

anthocyanins: water-soluble pigments located in cell vacuoles that impart red, purple, and blue colours to cells and tissues

apical meristem: region of cell division at the tip of shoots and roots

astrosclereid: a star-shaped sclereid with arms that are shorter and wider than those of a trichosclereid

autofluorescence: emission of fluorescence from unstained samples when viewed with short wavelength light; also called primary fluorescence

bicollateral vascular bundle: vascular bundle with phloem on two sides of the xylem

birefringence: property of crystalline materials that allows them to rotate the plane of polarized light so that they appear bright on a dark background in a polarizing microscope

bisexual flower: flower with both male and female structures

blade: expanded, flattened portion of a leaf

border cells: cells that are at the periphery of the root cap

bordered pit: a type of pit in which the secondary cell wall overarches the pit membrane

bract: a modified leaf-like organ, often associated with flowers or inflorescences

brachysclereid: isodiametric cell with a thick, lignified secondary cell wall; also called a stone cell

branched pit: pit with two or more canals; also called a ramiform pit

bulliform cells: enlarged epidermal cells probably involved in leaf folding or rolling and unrolling through loss and gain of water, respectively

bundle cap: a group of cells, usually fibres, located on the phloem side of a vascular bundle

bundle sheath: layer of cells surrounding a vascular bundle

calcium carbonate crystal: crystal composed of calcium carbonate; commonly in the form of a cystolith

calcium oxalate crystals: crystals of various forms composed of calcium oxalate

callose: a complex carbohydrate (β-1,3 glucan) associated with sieve areas in lateral walls and the pores in sieve plates (end walls) of sieve tube members; also found at sites of injury and as 'plugs' in growing pollen tubes

cambial zone: the vascular cambium and its recent

derivatives giving rise to secondary xylem and secondary phloem

carotene: yellow or orange pigment belonging to the carotenoid class of pigments

carotenoid pigments: lipid-soluble pigments consisting of carotenes and xanthophylls giving orange and yellow colours to cells and tissues; located in chromoplasts

Casparian band (strip): band-like structure composed of lignin and suberin found in the anticlinal (radial and transverse) walls of the exodermis and endodermis

cellulose: principal component of plant cell walls; complex carbohydrate (β-1,4 glucan) composed of polymerized glucose molecules

cell wall: outermost structure of plant cells surrounding the protoplast; a rigid layer consisting of cellulose, hemicellulose, pectic substances, proteins, and water

chlorenchyma: a modified type of parenchyma cells with an abundance of chloroplasts

chlorophyll: green pigment in chloroplasts; receptor of light energy in photosynthesis

chloroplast: coloured plastid containing chlorophyll and other pigments; site of photosynthesis

chromoplast: coloured (but not green) plastid containing carotenoid pigments

cladophyll: a branch (stem) that resembles a leaf

closed vascular bundle: a bundle in which there is no potential for the initiation of a vascular cambium; common in monocots

closed venation: arrangement of vasculature in the leaf blade characterized by the joining of minor veins

collateral vascular bundle: a vascular bundle with phloem only on one side of the xylem

collenchyma: tissue consisting of living cells with unevenly thickened primary cell walls that are usually unlignified; provides strength to growing as well as non-growing regions of organs

companion cell: cytoplasm-rich parenchyma cell associated with a sieve tube member in the phloem of angiosperms; one of the cells formed by the division of the sieve-tube precursor cell

compound starch grain: a combination of several simple starch grains within one amyloplast

cork: equivalent to phellem; tissue produced by the phellogen that provides protection against drying and pathogens for stems and roots; dead cells with suberin and wax in their walls

cork cambium: equivalent to phellogen; lateral (secondary) meristem that produces phellem and phelloderm

cortex: a primary tissue external to the vascular tissues in roots and stems

cotyledon: component of the embryo often involved in storing reserve substances

covering trichome: an outgrowth of the epidermis on leaves, stems, and reproductive organs that lack a secretory function

cuticle: a protective covering over epidermal cells of stems and leaves consisting of cutin and wax

cutin: a polymer that may be deposited within cell walls or on the surface of epidermal cells

cyclosis: another term for cytoplasmic streaming; bulk flow of cytoplasm within a cell

cystolith: a deposit of calcium carbonate around an ingrowth (peg) of cell wall; found in an enlarged cell called a lithocyst

cytoplasm: living portion of the cell

cytoplasmic streaming: movement of cytoplasm along channels within cells

diarch: within roots refers to the presence of two protoxylem poles

druse crystal: a compound calcium oxalate crystal with many small crystals projecting from its surface

elaioplast: lipid-storing plastid

embryo: young plant formed within a seed

endodermis: innermost layer of cortex in roots (and some stems) characterized by the development of a Casparian band and in some cases suberin lamellae and an additional secondary or tertiary cell wall

endothecium: the subepidermal wall layer in the anther with specialized secondary wall thickenings

ephedroid perforation plate: end wall of a vessel member that consists of a series of circular holes

epidermis: the outermost layer of cells in leaves, roots, stems and floral parts in primary growth

epifluorescence microscope: a compound microscope with a mercury vapour light source positioned above the objectives and equipped with special filters to produce short wavelengths of light

epithelial cells: secretory cells surrounding various ducts and canals

ergastic substance: a storage or waste product in plant cells

exine: outer wall of a pollen grain

exodermis: specialized outer layer(s) of roots with similar characteristics to endodermis, especially possessing a Casparian band

external phloem: in a bicollateral bundle, primary phloem located to the cortical side of the primary xylem

extrafloral nectary: nectary situated on a plant part other than a flower

fascicular vascular cambium: portion of vascular cambium originating within a vascular bundle

fibre: an elongated cell, dead at maturity, with a thick, lignified secondary cell wall; provides support for plant organs

filament: structure supporting an anther in flowers; component of a stamen

foraminate perforation plate: see ephedroid perforation plate

glandular trichome: an outgrowth of the epidermis on leaves, stems, and reproductive organs that have a secretory function; also called secretory trichomes

ground parenchyma: term used, for example, in stems in which there is no differentiation into cortex and pith

ground tissue: all primary tissues in an organ other than epidermis and vascular tisssues

guard cells: specialized cells in the epidermis of leaves and shoots that are able to swell and, thus, open a pore (the stoma)

guttation: loss of liquid derived from the xylem through structures on leaves called hydathodes

gynoecium: collective term for the carpels in a flower

hair: outgrowth of an epidermal cell

helical cell wall thickening: deposition of lignified secondary cell wall in a spiral pattern in xylem tracheary elements

hydathode: specialized structure within leaves through which xylem sap is expelled by root pressure

hydrophyte: a plant that either lives in water or must have an abundance of water for survival

hypocotyl: portion of embryo or young seedling between the cotyledon(s) and the radicle

hypodermis: one or more layers of cells underlying the epidermis that differ from adjacent cells

idioblast: a cell that differs from neighbouring cells by size, contents or wall structure

intercellular space: space between two or more cells; usually filled with air

interfascicular vascular cambium: vascular cambium that is initiated in parenchyma between adjacent vascular bundles in stems

internal phloem: primary phloem located to the inside of primary xylem

internode: region between two adjacent nodes on a stem

Kranz anatomy: wreath-like arrangement of mesophyll cells around the bundle sheath in leaves such as those of corn.

lacunar collenchyma: type of collenchyma in which the thickened primary cell wall is adjacent to intercellular spaces

lamellar collenchyma: type of collenchyma in which the thickened primary cell wall is deposited along the inner and outer tangential walls

lateral meristem: meristems that give rise to secondary tissues; vascular cambium and phellogen

lateral root: root that is initiated from another root

lateral sieve area: a portion of a sieve element wall containing small diameter pores through which protoplasts of adjacent cells are connected

latex: a fluid, often milky, contained within laticifers

laticifer: cell or a series of cells containing latex

leaf primordium: lateral outgrowth from shoot apical meristem that will differentiate into a leaf

lenticel: area within the periderm of roots and stems modified for air exchange; a raised area with broken cork cells

leucoplast: a colourless plastid, i.e., lacking pigments

lignin: complex polymer of phenyl propane subunits that gives additional strength to cell walls; common deposit in secondary cell walls

lipids: organic substances such as fats and oils that are insoluble in water

lithocyst: a specialized cell containing a cystolith

locule: cavity within the ovary containing ovules

lumen: space within a dead cell, surrounded by the cell wall

macrosclereid: elongated type of sclereid common in some seed coats; e.g., of legumes

major veins: large veins within a leaf

medullary bundles: additional vascular bundles located in the pith region of stems

meristem: a region of cells undergoing division from which primary tissues are formed

mesophyll: photosynthetic parenchyma cells in the leaf blade

metaxylem: part of primary xylem that matures after the protoxylem and beyond the zone of elongation

microsporangium (plural, microsporangia): in angiosperms, the anther locule and its walls in which microspores (pollen grains) are formed

middle lamella: pectin-rich layer that binds primary cell walls of adjacent cells together

midrib: parenchyma (sometimes collenchyma or sclerenchyma) associated with the midvein of leaves

midvein: the large central vein in some leaves

minor veins: smallest veins more or less embedded in the mesophyll of leaves

mitochondrion (plural, mitochondria): organelles involved in respiration and energy production

mucilage: carbohydrate secretion produced by root cap and epidermal cells

multiple epidermis: peripheral region of some stems, roots, and leaves consisting of more than one layer of cells; derived from a single cell layer in the apical meristems

nectary: a structure that secretes a substance, nectar, composed primarily of sugars but that also may contain amino acids and other compounds

netted venation: arrangement of veins in leaves resembling a net; most common in dicotyledonous species, also called reticulate venation

node: site on a stem from which leaves and axillary buds arise

nucleus (plural, nuclei): organelle containing chromatin and nucleoplasm

open vascular bundle: a type of vascular bundle that has the potential to initiate a vascular cambium; a feature of dicotyledonous species

open venation: type of venation in leaf blades in which minor veins end freely in the mesophyll

organ: structure composed of tissues

osteosclereid: bone-shaped sclereid; sometimes called an hour-glass cell; found in legume seed coats

ovary: part of flower containing ovules

ovule: structure within the ovary that develops into a seed

palisade mesophyll: elongated cells in leaves that have their long axis oriented perpendicular to the epidermis

paradermal section: section cut parallel to the surface of an organ

parallel venation: arrangement of veins in which the main veins are parallel with each other; common in leaves of monocotyledonous species

parenchyma: a tissue composed of parenchyma cells that are characterized by being alive and usually having only an unmodified primary cell wall

passage cell: cell in endodermis that has a Casparian strip but no suberin lamellae or a secondary cellulosic cell wall; the site of passage of water from the cortex to the stele

pectin: polysaccharide, mainly polygalacturonic acid, that is a principal component of primary cell walls

pedicel: stem-like structure subtending a flower within an inflorescence

peltate scale: multicellular trichome with a stalk and umbrella-shaped cap

perforation plate: end walls of vessel elements in the xylem that have one or several openings

pericarp: the fruit wall

periclinal cell division: division of a cell in which the new wall formed is parallel to the surface of the organ

pericycle: layer or layers of cells between the endodermis and vascular tissues in roots; outermost layer of vascular cylinder

periderm: complex tissue consisting of phellogen, phellem, and phelloderm providing protection for roots and stems subsequent to secondary growth

petiole: structure subtending the leaf blade

phellem: protective tissue (cork) formed by the phellogen; mature cells are dead with suberin and wax in their walls

phelloderm: tissue resembling cortex formed by the phellogen

phellogen: lateral meristem that forms phelloderm and phellem; outer bark

phi thickenings: lignified secondary wall deposits found in cortical cells of some roots

phloem: complex tissue involved in transport of sugars and other compounds

pit: a gap in the secondary cell wall of a cell

pith: central region of some stems and roots consisting primarily of parenchyma cells

plastid: general term for an organelle with a double membrane, internal membranes, and stroma

pollen grain: a microspore containing the male gametophyte that has a specialized wall

pollen sac: a cavity in anthers containing pollen grains

pollen tube: outgrowth from a pollen grain that grows down the style and delivers sperms to ovules

polyarch: term to designate the xylem of roots that consists of many protoxylem poles (strands)

polyphenols: complex aromatic compounds; may be deposited in cell walls, vacuoles, or the cytoplasm

P-protein: a specialized protein found in sieve tube elements; occurs as fibrils or bodies

primary cell wall: wall formed mainly while the cell is still increasing in size

primary fluorescence: emission of light by biological compounds when excited with an appropriate wavelength of light; also called autofluorescence

primary growth: tissue formed from the apical meristems of shoots and roots

primary phloem: phloem that differentiates from procambium; one of the primary tissues in organs

primary pit field: a thin area in the primary cell wall through which plasmodesmata pass

primary tissues: tissues derived from the apical meristems of shoots and roots

primary xylem: xylem that differentiates from procambium; one of the primary tissues in organs

prismatic crystal: crystal of calcium oxalate shaped like a prism

procambium: the primary meristematic tissue from which primary phloem and primary xylem are derived

proplastid: a small organelle (plastid) from which plastids develop

protein: a complex organic substance composed of amino acids

proteinoplast: plastid that stores protein

protoplast: the living unit of a single cell confined by the cell wall

protoxylem: the first mature primary xylem in an organ; becomes functional while organ is still elongating

protoxylem lacuna: large space within the xylem resulting from destruction of protoxylem tracheary elements

provascular tissue: a term sometimes used in place of procambium; the primary meristematic tissue from which primary phloem and primary xylem are derived

radicle: embryonic root; the first root to emerge during seed germination

ramiform pit: branched simple pit that occurs in secondary cell walls

raphide crystal: needle-like crystal of calcium oxalate; often occur in groups

ray: parenchyma tissue formed by ray initials in vascular cambium

resin duct: a canal lined by epithelial cells that secrete resin

reticulate cell wall thickening: a net-like pattern of secondary cell wall deposition in tracheary elements

reticulate venation: arrangement of veins resembling a net in a leaf

root cap: protective covering over the apical meristem of a root

root hair: extensions of an epidermal cell that provide an increased surface area for uptake of water and nutrients

root hair papilla: initial bulging of a root epidermal cell in the process of forming a root hair

scalariform cell wall thickening: deposition of lignified secondary cell wall in a ladder-like pattern in xylem tracheary elements

scalariform perforation plate: multiperforate plate in the end walls of xylem vessel members in which openings are separated by bands of wall in a ladder-like pattern

sclereid: type of sclerenchyma cell, usually not more than three times as long as wide; develops a lignified secondary cell wall; dead at maturity

sclerenchyma: support tissue consisting of cells with lignified secondary cell walls; includes both sclereids and fibres

sclerified bundle sheath: layer of cells with lignified walls surrounding a vascular bundle

secondary cortex: parenchymatous tissue formed in some monocotyledonous stems that have anomalous secondary growth

secondary phloem: phloem derived from derivatives of the vascular cambium

secondary tissues: tissues produced by the lateral meristems, i.e., vascular cambium and phellogen

secondary vascular bundles: additional vascular bundles added to stems of some monocotyledonous species following the completion of primary growth

secondary wall: wall deposited on the inside of the primary cell wall usually after cell growth is complete

secondary xylem: xylem derived from derivatives of the vascular cambium

secretory cavity: a space containing a secretory substance formed by the breakdown of cells

secretory duct: a tube-like structure containing a secretion produced by secretory cells (epithelial cells) lining the duct

secretory structure: any cell or group of cells in the plant body involved in producing a secretory product

seed coat: another term for testa, the outer protective layers of the seed

septate: divided by cross-walls as in septate fibres

shoot: part of the plant consisting of stems and leaves; includes flowers and fruits when present

sieve cell: sieve element with unspecialized sieve areas and a lack of sieve plates; found in the phloem of gymnopserms and seedless vascular plants

sieve elements: cells in the phloem involved in long-distance transport of sugars; further divided into sieve tube members and sieve cells

sieve plate: specialized wall containing pores at the ends of sieve tube members

sieve tube: two or more sieve tube members arranged end to end

sieve tube member: cell of the phloem involved in long-distance transport of sugars

simple perforation plate: single perforation in end wall of xylem vessel members

simple sieve plate: transverse wall consisting of one region of pores at end of a sieve tube member

simple starch grain: grain consisting of a single hilum around which starch is deposited

slime plug: an accumulation of P-protein against the sieve plate; P-protein is displaced as a result of wounding

sperm cell: male gamete; one fertilizes an egg cell to form a zygote

spongy mesophyll: parenchyma cells with large intercellular spaces in the leaf

stamen: the male part of the flower usually consisting of an anther and filament; responsible for pollen production

staminal column: structure formed by the fusion of filaments around the style within a flower

starch: a polysaccharide composed of glucose molecules; a main storage compound in plants

starch grain: the physical structure within an amyloplast or other plastid type composed of starch molecules

stigma: region at apex of a style on which pollen lands and germinates

stoma (pl. stomata): the opening (pore) in the epidermis of leaves and stems surrounded by two guard cells; sometimes called the stomatal pore

stomatal complex (stomatal apparatus): stoma, guard cells, and associated epidermal (subsidiary) cells

stomatal crypt: a depression in leaves of some species in which stomata are located

storage root: a modified root, usually by the production of large numbers of parenchyma cells, used for storage of starch, sugars, etc.

style: extension of the ovary bearing the stigma

suberin: polymer of fatty acids and phenolics found in phellem cells, the Casparian band, and suberin lamellae

suberin lamella (plural, suberin lamellae): thin layers of suberin deposited within cell walls

subsidiary cells: specialized epidermal cells found adjacent to guard cells

sub-stomatal chamber: the large air space immediately beneath a stoma

tannins: glycosides containing polyhydroxyphenols or their derivatives; they may be deposited in cell walls, vacuoles, or the cytoplasm

tapetum: a layer of cells in anthers providing substances for the growth of pollen grains

tepal: member of the perianth of a flower that may be either petal-like or sepal-like

testa: seed coat

tetrarch: roots with four protoxylem poles

tip growth: extension of structures such as root hairs and pollen tubes by addition of cell wall material to the apex of the cell

tissue: group of cells that perform a specific function

totipotent: term used to describe any cell or tissue that is capable of forming the whole plant

tracheary element: conducting cells in the xylem; a general term including tracheids and vessel members

tracheid: elongated cell with a lignified secondary cell wall with pits but no perforations; found in the xylem of all vascular plant groups; involved in transport of water and nutrients

transfusion tissue: in conifer leaves, the tissue consisting of tracheids and parenchyma cells surrounding the vascular tissues

transmitting tissue: tissue in the style of a flower along which or within which pollen tubes grow

transvacuolar strand: strand of cytoplasm that transverses vacuoles

triarch: term to designate roots with three protoxylem poles

trichome: commonly called a hair; outgrowth of an epidermal cell

trichosclereid: a branched sclereid, often with long, narrow arms

tylose: a plug resulting from the invasive growth of a parenchyma cell into the lumen of a vessel member

tylosome: an exodermal cell with wall ingrowths found in orchid roots

uniseriate epidermis: epidermis of any organ that consists of a single layer of cells

vacuolar pigments: water-soluble pigments, such as anthocyanins, that are deposited within vacuoles

vacuole: a region in the cell bound by a single membrane (tonoplast) and filled with water, water soluble materials, and insoluble precipitates

vascular bundle: a region consisting of xylem and phloem arranged on one radius

vascular cambium: the lateral meristem responsible for the formation of secondary xylem and secondary phloem

vascular cylinder: all tissues interior to the endodermis in roots

vein: vascular bundle in a leaf blade consisting of xylem and phloem

vein ending: the terminus of a vein in the mesophyll of a leaf, often consisting of enlarged tracheary elements

velamen: modified multiple epidermis of orchid roots

venation: the pattern of veins in leaves

vessel: two or more vessel members joined end to end for the transport of water and ions

vessel member: cell type in xylem with a lignified secondary cell wall and perforations in end walls for the long-distance transport of water and ions

wood: secondary xylem

xanthophyll: a yellowish carotenoid pigment

xylem: a complex tissue involved in long-distance transport of water and ions

young primary root: first root to emerge from a seed; technically known as the radicle

INDEX

A

accessory cell 141
Acer 28, 29
adventitious root 68, 72, 79, 141
aerenchyma 31, 65, 79, 93, 136, 141
aerial root viii, 76, 77, 141
Aesculus hippocastanum 117
African violet 27, 55, 117
Allium cepa 17, 68, 69
Alstroemeria 33, 34, 122, 123, 124, 137
Amaranthus 112
amyloplast vii, 18, 20, 141, 142, 148
angular collenchyma 33, 34, 35, 83, 84, 87, 93, 141
anther 13, 120, 121, 141, 143, 144, 148
anthocyanin 13, 27, 87, 99, 141, 149
apical meristem 76, 81, 82, 89, 141, 144, 147
Apium graveolens 31, 33, 34, 60
apple 73
Arabidopsis 131
Arabidopsis thaliana 105, 113, 114, 130, 131
Arachis hypogaea 40
Argyreia nervosa 88, 117
Aristolochia durior 28, 46, 95, 96
Asimina 47
Asparagus densiflorus 15, 16
asparagus fern 15, 16, 18
astrosclereid 40, 141
autofluorescence 72, 77, 91, 92, 141, 146
avocado 23, 38

B

banana 28, 29, 58, 59
basswood 47, 113, 114
bean 21, 24, 48, 50, 90, 102
beefsteak plant 87
beet 27
begonia 93, 101
Begonia rex 93, 101
bell peppers 18
Beta vulgaris 27
Betula 47
bicollateral vascular bundle 51, 83, 141
birch 47
Bird of Paradise 19
birefringence 40, 141
border cells 63, 141
bordered pit 45, 47, 141
brachysclereid 38, 40, 93, 141
branched pit 141
Brassavola nodosa 77
Brassica napus 57
bromeliad 101
bulliform cells 102, 110, 141
bundle cap 36, 51, 83, 91, 141

bundle sheath ix, 86, 102, 110, 111, 112, 113, 114, 141, 144, 147
buttercup 65, 66

C

calcium carbonate crystals vii, 26, 141
calcium oxalate crystal vii, 25, 104, 141, 143
callose 53, 132, 141
cambial zone 78, 89, 91, 141
Campsis radicans 96
Canadian pondweed 15, 16, 82
Canna generalis 31
canna lily 31
canola 57
Capsicum annuum 18, 19
carotene 18, 142
carotenoid pigments 18, 19, 20, 142, 149
carrot 7, 8, 20, 78
Casparian band (strip) 65, 67, 68, 69, 70, 71, 132, 142, 143, 148
Cattleya 77
Cattleya aurantiaca 77
celery 7, 31, 33, 34, 60
cellulose 36, 40, 68, 70, 142
cell wall 14, 26, 27, 31, 32, 33, 34, 35, 36, 37, 38, 40, 41, 42, 43, 45, 46, 47, 63, 65, 66, 68, 69, 70, 76, 77, 87, 91, 104, 133, 141, 142, 143, 144, 145, 146, 147, 148, 149
chlorenchyma 142
chlorophyll 15, 18, 133, 142
Chlorophytum comosum 85, 111
chloroplast 15, 16, 18, 19, 31, 41, 59, 76, 77, 83, 84, 99, 111, 117, 132, 133, 142
chromoplast vii, 18, 19, 20, 132, 142
chrysanthemum 55, 90, 91, 103, 113, 114
Chrysanthemum morifolium 55, 90, 91, 103, 113, 114
Citrus sinensis 60
cladophyll 15, 16, 18, 142
closed vascular bundle 85, 142
closed venation 113, 142
coleus 28, 42, 55, 81, 82, 103
Coleus blumei 28, 42, 55, 81, 103
collateral vascular bundle 83, 86, 142
collenchyma vii, 33, 34, 35, 83, 84, 85, 87, 93, 141, 142, 144
Commelina 14
common bean 21, 24
companion cell 51, 52, 86, 110, 142
compound starch grain 21, 78, 142
Cordyline 96
cork 74, 89, 91, 142, 144, 145
cork cambium 74, 89, 91, 142
corn 51, 63, 67, 68, 73, 74, 82, 86, 99, 102, 110, 111, 112, 113, 114
cortex 28, 31, 33, 38, 51, 58, 61, 65, 66, 67, 68, 69, 70, 71,

73, 74, 76, 77, 79, 83, 84, 85, 87, 88, 89, 91, 93, 96, 97, 117, 142, 143, 145, 147
cotyledon 23, 24, 130, 142, 143
covering trichome ix, 15, 88, 102, 103, 117, 142
crossed polars 22
cucumber 51, 53, 103
Cucumis sativus 51, 53
curly-leaved pondweed 93, 94, 115, 116
cuticle 1, 14, 106, 107, 133, 138, 142
cutin 133, 142
cyclosis 13, 142
cystolith 26, 141, 142, 144
cytoplasm 13, 14, 27, 31, 33, 58, 63, 70, 76, 133, 138, 139, 142, 146, 148, 149
cytoplasmic streaming 13, 14, 15, 17, 139, 142

D

Datura stramonium 32, 33, 34, 43, 103
Daucus carota 20, 78
desert privet 106
diarch 66, 142
Dracaena 96, 97
druse crystal 25, 78, 79, 108, 113, 114
Dutchman's pipe 28, 46, 95, 96
dwarf spike rush 93, 94

E

Easter lily 119, 120, 121, 122
Eastern white-cedar 73
Ecballium elaterium 103
Egeria 15, 18
Eichhornia crasipes 79
elaioplast 18, 20, 143
Eleocharis parvula 93, 94
Elodea 15, 16, 18, 82
Elodea canadensis 15, 16, 82
embryo ix, 130, 131, 141, 142, 143
endodermis viii, 65, 66, 67, 68, 69, 70, 71, 73, 74, 76, 106, 107, 133, 142, 143, 145, 149
endothecium 121, 143
Ephedra 45, 46, 47
ephedroid perforation plate 143
epidermis ix, 17, 26, 28, 33, 34, 67, 68, 69, 70, 71, 72, 76, 84, 85, 86, 88, 89, 91, 93, 99, 100, 102, 106, 107, 108, 110, 115, 116, 121, 138, 141, 142, 143, 145, 148, 149
epifluorescence microscope 15, 16, 28, 53, 67, 68, 91, 143
epithelial cells 58, 59, 60, 143, 147
ergastic substance vii, 13, 25, 143
Euphorbia splendens 61, 113
exine 120, 121, 143
exodermis viii, 67, 68, 69, 71, 76, 77, 133, 142, 143
external phloem 51, 83, 84, 143
extrafloral nectary 58, 143

F

fascicular vascular cambium 89, 90, 91, 95, 143
Festuca 63, 64, 73, 74

fibre viii, 36, 37, 39, 41, 43, 45, 46, 47, 51, 52, 58, 59, 83, 84, 85, 86, 87, 91, 95, 96, 141, 143, 147
Ficus elastica 26
filament 13, 120, 121, 143, 148
foraminate perforation plate 143
Fragaria ananassa 27

G

garden beet 27
geranium 25, 27, 55, 72, 91, 92, 113, 114, 117
glandular trichome(s) 55, 143
Glycine max 33, 34, 41, 74
grape 45, 46, 47, 117
ground parenchyma 120, 143
ground tissue 85, 86, 121, 143
guard cells 57, 99, 106, 107, 108, 109, 110, 111, 141, 143, 148
guttation 58, 143
Gymnocladus 47
gynoecium 120, 143
Gynura sarmentosa 103

H

hair viii, 13, 24, 55, 63, 64, 67, 143, 147, 148, 149
Haworthia 106, 107
Hedera helix 103
Helianthus annuus 23, 41, 42, 44, 51, 58, 59, 83
helical cell wall thickening 143
Hibiscus 106, 125, 126, 127, 128
Hibiscus rosa-sinensis 125, 126, 127, 128
horse-chestnut 117
Hoya carnosa 38
hydathode viii, 58, 143
hydrophyte 143
hypocotyl 143
hypodermis 72, 106, 107, 143

I

idioblast 25, 144
Impatiens 25, 129
intercellular space 31, 33, 35, 88, 106, 144, 148
interfascicular vascular cambium 89, 90, 95, 144
internal phloem 84, 144
internode viii, 38, 41, 42, 44, 51, 83, 86, 90, 91, 95, 144
Ipomoea batatas 21, 25, 78, 79
Ipomoea purpurea 46
Iresine herbstii 87
ivy 103

J

Jimson weed 32, 33, 34, 43, 103

K

Kalanchoe 99
Kentucky coffee tree 47
Kranz anatomy 110, 144

L

lacuna 86, 116
lacunae 31, 93, 94, 110, 115
lacunar 33, 34, 35, 144
lacunar collenchyma 33, 35, 144
lamellar collenchyma 33, 34, 35, 85, 144
lateral meristem 89, 91, 96, 144, 146, 147, 149
lateral root viii, 64, 73, 74, 76, 144
lateral sieve area 53, 144
latex 58, 61, 144
laticifer viii, 61, 144
leaf primordium 144
lenticel 93, 144
leucoplast 18, 20, 144
lignin 1, 32, 36, 41, 42, 66, 70, 72, 77, 104, 132, 133, 142, 144
Lilium longiflorum 119, 120, 121
lipid vii, 18, 23, 91, 132, 133, 138, 142, 143, 144
lipid staining vii, 23
Liriodendron tulipifera 45
lithocyst 26, 142, 144
locule 120, 144
lumen 46, 60, 144, 149
Lythrum salicaria 79

M

macrosclereid 40, 144
major veins 144
Malus pumila 73
maple 28, 29
marigold 117
medullary bundles 144
Mentha 55
meristem 76, 81, 82, 89, 91, 96, 132, 141, 142, 144, 146, 147, 149
mesophyll 36, 58, 59, 101, 106, 107, 108, 109, 111, 112, 115, 144, 145, 148, 149
metaxylem 41, 42, 43, 65, 66, 67, 83, 84, 86, 110, 144
microscopes
  compound 1
  dissecting 1
microsporangium (plural, microsporangia) 120, 144
middle lamella 144
midrib 31, 105, 109, 144
midvein 108, 109, 144
minor vein 102, 111, 113, 142, 144, 145
mints 55
mitochondria 15, 16, 133, 145
Monstera deliciosa 40
morning glory 46
mother of thousands 102
mucilage 25, 63, 145
multiple epidermis 145, 149
Musa 28, 29, 58, 59
Myriophyllum 93, 94, 115, 116

N

Nasturtium 104
nectary 57, 58, 143, 145
Nerium oleander 108, 109
netted venation 145
node 83, 144, 145
nuclei 1, 15, 16, 17, 58, 132, 133, 138, 139, 145
nucleus 13, 15, 16, 17, 31, 33, 138, 145
Nymphaea odorata 40, 115, 116

O

oak 28, 39, 45
oleander 108, 109
onion 17, 68, 69, 138, 139
open vascular bundle 145
open venation 113, 145
orange 60
orchid(s) 76, 77
osteosclereid 40, 145
ovary 120, 144, 145, 148
ovule ix, 120, 130, 131, 144, 145

P

palisade mesophyll 106, 108, 115, 145
pansy 19
paradermal section 60, 69, 71
parallel venation 113, 145
parenchyma vii, 23, 27, 28, 31, 33, 35, 38, 39, 40, 41, 43, 45, 46, 51, 52, 59, 61, 65, 78, 79, 83, 84, 87, 89, 93, 94, 95, 106, 108, 109, 110, 115, 116, 117, 120, 121, 141, 142, 143, 144, 145, 146, 148, 149
passage cell 69, 73, 76, 77, 145
Passiflora caerulea 58
passion flower 58
pawpaw 47
pea 40, 74
peace lily 117
peanut 40
pear 38
pearl plant 106, 107
pectin 32, 40, 41, 87, 133, 144, 145
Pelargonium hortorum 25, 27, 55, 72, 91, 92, 113, 114, 117
peltate scale ix, 101, 102, 145
Peperomia magnoliaefolia 106
Persea americana 23, 38
Peruvian lily 122
perforation plate 43, 45, 46, 47, 143, 145, 147
pericarp 145
periclinal cell division 91, 145
pericycle 65, 66, 67, 68, 74, 78, 145
periderm viii, 74, 89, 91, 93, 144, 145
petiole ix, 15, 28, 29, 31, 33, 34, 55, 58, 60, 115, 116, 117, 145
Phalaenopsis 77
Phaseolus vulgaris 21, 24, 48, 50, 90, 102
phellem 20, 74, 78, 89, 91, 92, 93, 142, 145, 146, 148
phelloderm 74, 89, 91, 92, 142, 145, 146
phellogen 74, 79, 89, 91, 92, 93, 142, 144, 145, 146, 147

phi thickenings viii, 72, 73, 146
phloem viii, 36, 41, 48, 51, 52, 61, 65, 66, 67, 68, 74, 77, 78, 79, 83, 84, 85, 86, 87, 88, 89, 90, 91, 93, 94, 95, 96, 97, 102, 106, 107, 108, 109, 110, 111, 113, 115, 116, 141, 142, 143, 144, 146, 147, 149
pigweed 112
Pilea cadierei 26, 58, 103
pine 28, 29, 45, 47, 58, 59, 106, 107
Pinus 28, 29, 45, 47, 58, 59, 76, 106, 107
Pinus strobus 45, 58, 59
Pinus sylvestris 29, 58, 59, 107
Pisum sativum 40, 74
pit 36, 37, 38, 45, 46, 47, 141, 146, 148
pith 31, 38, 41, 61, 65, 67, 68, 76, 83, 84, 85, 88, 93, 95, 96, 117, 143, 144, 146
plastid vii, 18, 19, 20, 31, 141, 142, 143, 144, 146, 148
Potamogeton crispus 93, 94, 115, 116
potato 7, 8, 20, 21, 91, 92
polarizing microscopy 19, 20, 21, 22, 25, 26, 36, 40, 46
pollen grain 120, 121, 141, 143, 144, 146, 148
pollen sac 120, 121, 146
polyarch 68, 146
polyphenol vii, 27, 133, 146
P-protein 53, 146, 148
primary cell wall 31, 33, 35, 37, 141, 142, 144, 145, 146, 147
primary fluorescence 70, 91, 141, 146
primary growth 89, 95, 143, 146, 147
primary phloem 41, 51, 52, 65, 67, 68, 83, 85, 86, 88, 89, 91, 95, 108, 110, 143, 144, 146
primary pit field 146
primary tissues viii, 65, 66, 67, 68, 69, 74, 75, 83, 89, 99, 142, 143, 144, 146
primary xylem 36, 41, 42, 43, 51, 65, 66, 67, 68, 69, 83, 84, 85, 86, 88, 89, 91, 95, 96, 110, 143, 144, 146
prismatic crystal 25, 40, 113, 114, 146
procambium 43, 52, 146
proplastid 18, 146
protein vii, 18, 20, 21, 23, 24, 53, 70, 132, 142, 146, 148
proteinoplast 18, 20, 146
protein staining vii, 23, 24
protoplast 33, 52, 69, 138, 142, 144
protoxylem 41, 42, 43, 65, 66, 67, 68, 83, 84, 86, 93, 94, 110, 115, 142, 144, 146, 148, 149
protoxylem lacuna 86, 93, 94, 110, 115, 146
provascular tissue 43, 52, 146
purple loosestrife 79
Pyrus communis 38

Q

Quercus 28, 39, 45

R

radicle 63, 64, 143, 146, 149
ramiform pit 38, 141, 146
Ranunculus 65
Ranunculus flabellaris 65, 66

raphide 25, 146
raphide crystal 25
ray 78, 95, 96, 146
resin duct 58, 59, 106, 107, 147
reticulate cell wall thickening 147
reticulate venation 113, 114, 145, 147
root cap 63, 74, 141, 145, 147
root hair viii, 13, 63, 64, 67, 147, 148
root hair papilla 147
Rose of China 125
rubber plant 26

S

sage 24, 55, 56
sail plant 117
Saintpaulia ionantha 27, 55, 117
Salix 47
Salvia officinalis 24, 55, 56
Sansevieria zeylanica 36
Saxifraga stolonifera 102
scalariform cell wall thickening 147
scalariform perforation plate 45, 147
Schefflera actinophylla 117
sclereid viii, 36, 37, 38, 40, 93, 94, 115, 116, 141, 144, 145, 147, 149
sclerenchyma viii, 36, 37, 87, 144, 147
sclerified bundle sheath 86, 147
Scots pine 58, 59, 107
secondary cortex 89, 147
secondary phloem 52, 74, 78, 89, 90, 91, 96, 142, 147, 149
secondary tissues viii, 78, 89, 144, 147
secondary vascular bundles 96, 97, 147
secondary wall 42, 44, 76, 143, 146, 147
secondary xylem 41, 43, 45, 74, 78, 79, 84, 87, 89, 90, 91, 95, 96, 142, 147, 149
secretory cavity 60, 147
secretory duct viii, 58, 59, 60, 147
secretory structure viii, 55, 58, 147
seed coat 37, 39, 40, 144, 145, 147, 148
shoot viii, 15, 41, 42, 44, 50, 51, 64, 65, 70, 81, 82, 83, 90, 95, 96, 141, 143, 144, 146, 147
sieve cell 52, 147
sieve element 52, 53, 144, 147
sieve plate 51, 52, 53, 141, 147, 148
sieve tube 51, 52, 53, 83, 84, 86, 110, 141, 142, 146, 147, 148
sieve tube member 53, 142, 148
simple perforation plate 45, 147
simple sieve plate 51, 53, 148
simple starch grain 20, 142, 148
slime plug 52, 53, 148
snake plant 36
Solanum lycopersicum 15, 19, 65, 66, 74, 83, 84
Solanum tuberosum 8, 20, 21, 91, 92
soybean 33, 34, 41, 74
Spathiphyllum wallisii 117
spider plant 85, 111
split-leaf philodendron 40

spongy mesophyll 106, 108, 109, 115, 148
squirting cucumber 103
stamen 13, 57, 119, 120, 141, 143, 148
starch ix, 1, 18, 20, 21, 23, 61, 78, 79, 87, 110, 112, 132, 133, 141, 142, 148
starch grain 18, 20, 21, 61, 78, 79, 87, 132, 133, 142, 148
star window plant 106
stigma 119, 120, 148
stoma 57, 58, 101, 103, 104, 106, 108, 120, 143, 148
stomatal apparatus (stomatal complex) 99, 101, 108, 121, 148
stomatal crypt 108, 109, 148
storage root viii, 78, 148
strawberry 27
Strelitzia reginae 19
style 119, 120, 148, 149
sub-stomatal chamber 107, 110, 111, 148
suberin 1, 26, 65, 66, 69, 70, 71, 91, 92, 132, 133, 142, 143, 145, 148
suberin lamellae 65, 69, 70, 71, 143, 145, 148
subsidiary cells 99, 107, 110, 111, 141, 148
sunflower 23, 41, 42, 44, 51, 58, 59, 83
sweet potato 21, 25, 78, 79

T

Tagetes erecta 117
tannins vii, 27, 28, 29, 58, 73, 76, 133, 148
tapetum 121, 148
testa 40, 147, 148
tetrarch 65, 66, 74, 148
Thuja occidentalis 73
Tilia 47, 113, 114
Tilia americana 113, 114
tip growth 63, 148
tomato 15, 65, 66, 74, 83, 84
totipotent 33, 148
tracheary element viii, 31, 41, 42, 43, 44, 45, 46, 50, 58, 65, 66, 77, 79, 87, 102, 110, 113, 137, 141, 143, 146, 147, 148, 149
tracheid 43, 45, 47, 137, 148
Tradescantia 13, 14, 25, 99
Tradescantia pallida 13, 14, 99
transfusion tissue 106, 107, 148
transmitting tissue 120, 149
transvacuolar strand 13, 14, 15, 17, 63, 149
triarch 65, 66, 149
trichome viii, ix, 13, 14, 15, 16, 24, 55, 56, 57, 81, 88, 101, 102, 103, 104, 106, 108, 110, 113, 114, 117, 142, 143, 145, 149
trichosclereid 40, 115, 141, 149
Triticum vulgare 99
trumpet vine 96
Tulip tree 45
tylose 46, 149
tylosome 76, 149

U

umbrella tree 117
uniseriate epidermis 67, 149

V

vacuolar pigment vii, 27, 149
vacuole 13, 14, 27, 28, 31, 33, 139, 141, 146, 148, 149
vascular bundle 41, 51, 83, 85, 86, 87, 88, 89, 90, 93, 94, 95, 96, 97, 106, 110, 117, 120, 121, 141, 142, 143, 144, 145, 147, 149
vascular cambium 43, 52, 74, 79, 84, 85, 89, 90, 91, 95, 96, 141, 142, 143, 144, 145, 146, 147, 149
vascular cylinder 65, 66, 67, 68, 69, 70, 73, 76, 77, 145, 149
vein 102, 104, 105, 106, 111, 112, 113, 114, 115, 120, 133, 142, 144, 145, 147, 149
vein ending 113, 149
velamen 76, 77, 149
velvet plant 103
venation 104, 113, 114, 142, 145, 147, 149
vessel 39, 43, 45, 46, 47, 67, 76, 78, 137, 143, 145, 147, 148, 149
vessel member 39, 43, 45, 46, 47, 143, 147, 148, 149
Viola odorata 19
Vitis 45, 46, 47, 117
Vitis vinifera 45, 117

W

wandering jew 13
wandering sailor 13
water hyacinth 79
water lily 40, 115, 116
water-milfoil 93, 94
wax plant 38
wheat 99
white pine 59
willow 47
wood 39, 45, 46, 149
wooly morning glory 88, 117

X

xanthophyll 18, 142, 149
xylem 41, 50, 51, 58, 65, 66, 67, 68, 69, 70, 74, 77, 78, 79, 83, 84, 85, 86, 87, 88, 89, 90, 91, 93, 94, 95, 96, 97, 102, 106, 107, 108, 109, 110, 111, 113, 115, 116, 141, 142, 143, 144, 145, 146, 147, 148, 149

Y

young primary root 63, 149

Z

Zea mays 51, 63, 67, 68, 73, 74, 82, 86, 102, 110, 111, 112, 113, 114